P9-EML-204

The PARADOX *of* EVOLUTION

The PARADOX *of* EVOLUTION

The Strange Relationship between NATURAL SELECTION *and* REPRODUCTION

STEPHEN ROTHMAN

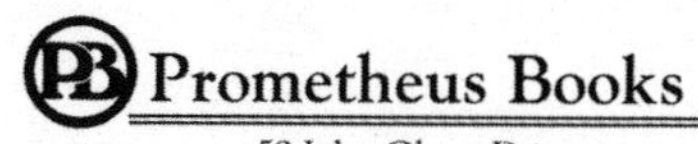

59 John Glenn Drive
Amherst, New York 14228

Published 2015 by Prometheus Books

The Paradox of Evolution: The Strange Relationship between Natural Selection and Reproduction.

Cover image © Bigstock
Cover design by Nicole Sommer-Lecht

Inquiries should be addressed to

Prometheus Books
59 John Glenn Drive
Amherst, New York 14228
VOICE: 716–691–0133
FAX: 716–691–0137
WWW.PROMETHEUSBOOKS.COM

19 18 17 16   5 4 3 2

Library of Congress Cataloging-in-Publication Data

Rothman, S. S. (Stephen S.), author.
The paradox of evolution : the strange relationship between natural selection and reproduction / by Stephen Rothman.
p. ; cm.
Includes bibliographical references and index.
ISBN 978-1-63388-072-6 (paperback) — ISBN 978-1-63388-073-3 (e-book)
I. Title.
[DNLM: 1. Selection, Genetic—Popular Works. 2. Biological Evolution—Popular Works. 3. Reproduction—Popular Works. QU 475]

QH390
572.8'38—dc23

2015026673

Printed in the United States of America

To Doreen, mother of my children and grandmother of theirs

CONTENTS

SECTION III—SEXUAL SELECTION

SECTION IV—PURPOSE

SECTION V—PURPOSE AND REPRODUCTION

One day as he was walking on the road, he (the sage Honi HaM'agel, Honi the Circle Maker) saw a man planting a carob tree. He asked him, "How long will it take this tree to bear fruit?" The man replied, "Seventy years." He asked, "Are you quite sure you will live another seventy years to eat its fruit?" He replied, "I myself found fully grown carob trees in the world; as my forbears planted for me, so I am planting for my children."

Translated from the Talmud, Tractate Ta'anit 23a

PREFACE

When I retired from the University of California–San Francisco after some forty years as a professor and research scientist, I decided to explore certain questions about the limitations of the reductionist scientific enterprise in biology that had been of interest to me for a long time, but that other obligations, the vagaries of university life, and my life otherwise had not permitted me to pursue.

The Paradox of Evolution is the third in a series on the subject. The first, *Lessons from the Living Cell*,[1] was an evaluation of reductionism as a research strategy. The second, *Life beyond Molecules and Genes*,[2] was about reductionism as a means of explaining life, of determining whether something is living. The current book discusses it as an instrument of description for evolution.

The questions raised in these books are unusual in that they concern points of view that many think are well settled, but about which I found myself skeptical, unconvinced, or at least uneasy. In writing about them, I have done my best to avoid asserting what cannot be argued with reason, not to believe as a matter of preconception what cannot be proven, and to confront rather than ignore what I don't understand or find disquieting.

With this mindset, *The Paradox of Evolution* is about my personal apprehensions as a committed Darwinian about the theory of evolution. In particular, it is about a serious problem that the reproductive features of life pose for Darwin's theory. It asks why we have children or, more pointedly, "Why in God's (or Darwin's) name do we have children?"

In light of this question the analysis that follows makes the revolutionary claim that the evolution of life's complex and diverse reproductive mechanisms is *not* the consequence of natural selection. The basis for the claim is outlined, and an alternative explanation for the evolution of life's reproductive features is offered. This not only raises obvious questions about the ubiquity of Darwin's theory; it exposes the deepest question possible about life's nature . . . its reason for being. What, it asks, accounts for the evolution of the mechanisms of reproduction? Not why it is necessary for life's evolution and continuance, without it there would be neither, but rather what in the natural world provoked its development?

As you shall see, it is not the details but the cogency of broad arguments that matter here. I believe that they are convincing, enlightening, and even heartening. Needless to say, you will make your own judgment. If nothing more I hope that you find what I have to say worthy of thought.

PROLOGUE

It probably will come as no revelation that sex is paradoxical. It can be a matter of attraction or repulsion; it can be exciting or routine, loving or aggressive, a simple bodily activity or a magical one, and it can change from one to the other with a word, a glance, or a touch. As for its product, our progeny, there is the pain of childbirth, the wonder of new life, and all the joy and suffering that follows raising a child. Given such contradictions, it should come as no surprise that the evolution of sexual reproduction, the means of producing offspring, presents a considerable scientific paradox.

What follows is an examination of one aspect of this paradox in a little-noted internal contradiction in Charles Darwin's and Alfred Russel Wallace's theory of biological evolution. The contradiction concerns a vitally important feature of the relationship between the theory's two essential elements, natural selection and reproduction—in particular, the role of natural selection in the evolution of life's reproductive features.

In offering an answer to one of humankind's most perplexing and important questions about nature, the implications of which go far beyond science, Darwin's theory has had a disputatious history. It seemed to pit science against religion, reason against belief, and was used to argue for or against particular attitudes about the natural state of humans and the appropriateness of various religious, political, and social views. Even with the press of noble intentions, when science and such judgment-laden ways of thinking collide, the truth is the first victim, and Darwin's theory has been no exception. Extreme opinions have often dominated the dialogue about the theory over its 150-year history. In the minds of many it freed humans from religious dogma. In great part as a consequence of this view, things have been topsy-turvy. The skeptics have in the main been religious believers, not scientists, and the acolytes, the scientifically inclined.

Though there was an ebb and flow of support for Darwin's theory for many years after publication of *On the Origin of Species by Means of Natural Selection* in 1859, it seemed that for all intents and purposes its acceptance in the scientific community was complete by the end of the 1930s with the introduction of the modern synthesis that explained evolution in genetic terms. The modern synthesis resolved what appeared to be a fundamental conflict between Gregor Mendel's genetic view of inheritance and Darwin's natural selection.

Since then, despite periodic questions about one or another detail of the theory, it is fair to say that Darwin's theory has become settled science, a central paradigm of biology. Though commonly thought of as an established fact of nature proven by incontrovertible evidence, no longer a suspect theory, some scientists, philosophers, and laypersons not only think it unproven but see it as little more than an unsubstantiated, even an unjustified hypothesis.

Given these circumstances and the theory's special place among scientific ideas, publicly contemplating problems with it and the evidence that supports it, however sound and dispassionate the analysis, is a thorny business. It may be perceived by believers as nothing less than treason, as giving succor to the "enemy." In the current case, the situation is not helped by the fact that the paradox in question is neither subtle nor hidden. If it is not self-evident, it is at least fairly obvious. Indeed, it is hard not to wonder why so many smart, even brilliant individuals who have thought deeply about the subject, beginning with Darwin and Wallace themselves, failed to consider it. Despite being conspicuous, it seems to have gone almost unnoticed, even disregarded.

In her 1991 book *The Ant and the Peacock: Altruism and Sexual Selection from Darwin to Today* Helena Cronin points out that one aspect of the topic, Darwin's theory of sexual selection, was not mentioned in most major books on evolution, and when it was discussion was perfunctory.[1] This even though Darwin had written a major book on the subject (*The Descent of Man and Selection in Relation to Sex*, 1871), and it was one of the few significant points of contention between Wallace and Darwin.[2]

When I first looked into the subject I presumed that in all likelihood the paradox was merely ostensible and that as I delved deeper my qualms would be dispelled. But this is not what happened. Things were not so easily resolved. This raised the difficult *nonscientific* question alluded to above.

Given its preeminence, if there really was a significant problem with Darwin's theory, how could it be successfully articulated?

As a matter of science, things were clear. The difficulty could not be ignored. It was not acceptable to acquiesce to beliefs about which one had doubts. A scientist's most elevated obligation is skepticism, not merely toward the new and the out of the ordinary but toward the common view; that held with the most certainty and most dearly embraced. In the absence of skepticism, there is no science, just dogma.

The history of science teaches us that truth in science as in life otherwise is not merely a matter of taking a vote. The common view is very often wrong, and sometimes in great error. Add to this the fact that no scientific theory, not even Darwin's, can ever be proven true beyond doubt. Scientific ideas, as the philosopher Karl Popper explained, however well established, however well regarded, are forever vulnerable to being shown false. Constantly probing their claims, perpetually seeking their falsification, is science of the highest order. Nor does logic allow us to take the absence of an acceptable well-developed alternative to a scientific theory, as has been the case for Darwin's theory, as proof of that theory's correctness. It may be provably incorrect and the absence of an alternative no more than a failure of the human imagination.

Even if we agree about all this, the fact cannot be avoided that considering a serious difficulty with such an important theory faces great obstacles. It is important to recognize at the outset that contemplating problems with the theory, as I do here, does not in any way denigrate it or its authors' great contribution. However the paradox is resolved, and I hope that I am able to point the way, it is essential to express heartfelt appreciation to Darwin and Wallace, not only for their acuteness of thought and observation, the breadth and importance of their insights, and their brilliant innovation, but first and foremost for their creative courage.[3] But, as I am sure they would agree, in the end, we must reserve our most profound respect not for a theory but for the breathtaking and humbling nature of life.

If you hope to be regaled by exotic examples of nature's marvels in what follows, I am afraid this book will disappoint you. There are many superb books, beginning with Darwin's, that do this far better than I ever could. But more to the point, this is not my purpose. I am not interested in cataloging examples of the theory's purview but in exploring its logical basis. Indeed, I have tried to avoid what evolutionary biologist Stephen Jay Gould, with a tip of the hat to Rudyard Kipling, called "Just So Stories,"[4] fanciful, appealing, often seemingly sensible explanations for events of evolution whose basis is, in fact, unknown. Still, it is hard to totally avoid flights of imagination in thinking about occurrences in the shadowed past, and I hope that I have not abused that authorial privilege. The task is to do our best to stick to what nature, not our imagination, tells us, using our deductive powers to gain understanding.

The theory of evolution's most important concept, natural selection, is central to what follows. Despite the idea's storied place in biology, it has had a troubled history. As suggested, even today some experts reject the concept as hopelessly flawed, while others insist that it is critical to any serious understanding of evolution. At least at first glance, the term's meaning is straightforward. Natural selection is any interaction, major or minor, between the environment and an organism that is potentially harmful or damaging to it in the face of which some creatures fare better than others. As Darwin first announced in *On the Origin of Species*, natural selection is "the preservation of favourable variations [in organisms] and the rejection of injurious ones [by nature]."[5] This said, even in Darwin's original descriptions, let alone in those advanced since, the meaning of the term has often been ambiguous and confusing. In chapter 4 I attempt to shed some light on this critical issue.

In a great many places I refer to "Darwin's theory," not "Darwin's and Wallace's theory," or to Darwin's view, not Darwin's and Wallace's view, when many of the ideas were independently imagined by both men. Though this is common enough, it is graceless and discomforting. Nonetheless, I have accepted it as necessary for literary simplicity, as well as cogency. It helps to avoid the awkwardness of having to mention both men's names every time their contributions are noted, which is constantly, and in mentioning them parsing their relative contributions. If I were to specify and compare their contributions at each and every juncture in the book, that would become its subject. It is important to appreciate that my choice is not a value judgment. For both good and bad reasons the theory is widely spoken of as Darwin's, not Wallace's.

SECTION I

DARWIN'S THEORY

Chapter 1

TWO MYSTERIES

A Story of Nature and Science

> *False facts are highly injurious to the progress of science, for they often endure long; but false views, if supported by some evidence, do little harm; and when this is done, one path towards error is closed and the road to truth is often at the same time opened.*
>
> Charles Darwin, *The Descent of Man*, 1871

This is a book about two mysteries, one of nature, another of science. A mystery of nature is of course something that we do not understand about nature. A mystery of science is something unknown or unrevealed about the thinking and methods of its practitioners. Both mysteries have to do with Darwin's theory of evolution by means of natural selection and its relationship to the reproductive mechanisms of life.

The mystery of nature is how reproductive mechanisms evolved. The mystery of science is why there has been so little discussion of the basis for the widespread belief that natural selection is responsible. When so much about Darwin's theory has been and continues to be the subject of far-reaching discussion and vigorous debate in both the scientific and popular literature, why has such an important topic been overlooked? One would have expected it not just to be present but to be prominent. Yet, as best I can tell, it has never been seriously considered, let alone in any depth.

NO DISCUSSION?

But before we go running off to look for an explanation, am I correct? Has the subject really been overlooked? Is there a genuine mystery here or am I simply mistaken? At first glance I appear to be incorrect and in no small measure at that. Without doubt, reproduction *has* been the subject of discussion in the scientific literature on evolution—substantial discussion. Indeed whole fields have been devoted to it.

As Darwin described evolution, it is a play with two equally important and interdependent acts. In the first, individual organisms, alone or in groups, face environmental threats of all sorts. Some survive the challenge, while others succumb. This is *survival* of the fittest (or fitter) and it is the work of the nonreproductive or *somatic* features of life.[1] As act one unfolds, we see natural selection working its will, determining the fate of organisms and species.

As dramatic as this is, it is merely the prologue to the second act. In it, we learn of sex and love, of generations to come. It is not about survival of the fittest but the *fecundity* of the fittest, about the work of reproduction.[2] And so evolution, indeed life itself, is comprised of two key elements—survival and reproduction or, we might say, food and sex. Evolution requires that organisms both survive to sexual maturity *and* then reproduce. Natural selection is only the agent of evolution *if* reproduction produces the next generation. It does not stand on its own. If an organism does not have progeny, it plays no role in evolution. This understanding is central to Darwin's theory and has been considered at some length, and in this sense it can be fairly said that not only has there been a discussion of reproduction; there has been an unmistakable appreciation of its central role in evolution. It has hardly been absent.

It is also understood that the role of reproduction in evolution is not just procreative. Even though Darwin did not know of Mendel's genetic ideas, from his knowledge of animal domestication he understood that reproduction was a critical source of the variety needed for natural selection to work its wonders. In this case, too, its role has not been ignored. To the contrary, it has been a critical concern of both classical and modern genetics. Perhaps the most relevant aspect of genetics has been population or evolutionary genetics, a primarily mathematical pursuit that considers the relationship between sex and evolution in populations of organisms. We will take a brief look at this from the standpoint of natural selection in chapter 4.[3]

Beyond this, there is a distinctive selection that is associated with reproduction whose purpose is to obtain or choose a mate. Not only did organisms have to survive life's challenges, that is, not only did they have to survive natural selection, but if they did and reached sexual maturity, they became subject to another kind of selection, a sexual selection. As said, if an individual is to play a role in evolution, it has to successfully mate. This is described in Darwin's theory of sexual selection. Although until recently this theory has been widely ignored, a consideration of the manifold ways and byways of obtaining and choosing mates certainly has not.

NATURAL SELECTION AS THE AGENT OF REPRODUCTIVE EVOLUTION

Why then do I say that the subject has been given short shrift? The reason is that nothing I have said is about how the structures, mechanisms, processes, and adaptations of reproduction came into being, how *they* evolved. A knowledgeable reader might grumble that this has also been considered, at least to the degree necessary. In view of the fact that the theory of evolution by means of natural selection has been the focus of extensive, not to say enduring, discussion, reproductive traits, a subset of life's features, have inescapably been part of that discussion even if they have not been specifically taken into account. If we claim and offer evidence to support the view that natural selection serves as the basis for life's evolution, then both the claim and the evidence apply to life's reproductive features no less than to its somatic features. We do not need to single out the many different characteristics of reproduction to know how they came about.

Confidence in this conclusion is based on three very broad suppositions, though we could just as well call them three well-founded understandings. The first is that there is an *identity of kind* between life's somatic and reproductive features. However disparate otherwise, they are things of the same sort in physical and chemical terms and as a result their properties in this regard are necessarily of the same general type. The second is that because they are of the same type they came into being as the result of the *same agency*, and that agency is natural selection. And finally, natural selection *applies generally*, to all of life's characteristics. There is no basis or foundation for excluding reproductive traits from its agency and consequences.

If these propositions are true, what more is there to consider? Comparing and contrasting the evolution of life's somatic and reproductive features either as a general matter or in detail would serve no purpose. From an evolutionary point of view, whatever the differences, they are inconsequential.

And yet is it acceptable to say that a subject has been fully or even adequately considered, if it is not specifically taken into account, point by point, element by element, instead simply declaring it to be an example of something else, something more general? However reasonable, can we really, at least as a matter of science, rely on these three suppositions to conclude that natural selection is the agency responsible for the evolution of life's reproductive features without considering the particular circumstances of their evolution? Aren't we obliged to address the proposition that natural selection is that agency directly? However easily accommodated and seemingly sound our a priori assumptions may be, in the absence of direct scrutiny, they remain unfounded.

The desire to amalgamate the evolution of the somatic and reproductive features of life under the common guidance of natural selection is certainly understandable, nevertheless, however desirable and however reasonable, a conjecture is not scientific proof. Dating back to Darwin this elision between what is assumed and what is known has taken place. Though Darwin and those who followed understood that reproduction plays its own unique role in evolution, other than in sexual selection, it was thought that the evolution of its features came about in exactly the same way, the consequence of the same forces as the other (somatic) features of living things.

NO OTHER CHOICE

There is one other argument to be made against the claim that the role of natural selection in the evolution of life's reproductive traits has *not* been adequately explored. It is that no plausible scientific alternative exists. That natural selection is the agency responsible for the evolution for life's reproductive features is not only reasonable; it is science's only credible explanation. On what other basis could they have evolved? Without an alternative to consider, what purpose is served by questioning natural selection's dominion?

But as noted in the prologue, science cannot be satisfied with such a position. It cannot be content with the inability to conceive of an alternative as proof of the truth of a proposition. After all, as said, the inability may simply be due to a lack of imagination. The history of science is littered with understandings for which alternative accounts at first seemed impossible, only to find in the fullness of time that not only did an alternative exist; it provided a more satisfactory explanation. However ignorant we may be, we cannot forswear a critical consideration of the relationship between natural selection and the evolution of the reproductive features of life merely because we lack an alternative.

It is with this understanding in mind that this volume examines the proposition that natural selection is the agency responsible for the evolution of the reproductive features of life. As we shall see, not only is there is a problem with the claim; it also poses what appears to be a massive contradiction.

Chapter 2

A MOST EXTRAORDINARY THEORY

The Remarkable Impact of a Scientific Theory on Society

> *Depend upon it, you have earned the lasting gratitude of all thoughtful men. And as to the curs which will bark and yelp, you must recollect that some of your friends, at any rate, are endowed with an amount of combativeness which (though you have often and justly rebuked it) may stand you in good stead.*
>
> *I am sharpening up my claws and beak in readiness.*
>
> Thomas Henry Huxley in a letter to Darwin outlining support for (as well as disagreements with) Darwin's theory, November 23, 1859

I am a Darwinian! My credentials rest on at least one book. *Life beyond Molecules and Genes* not only depends on Charles Darwin and Alfred Russel Wallace's theory of evolution by means of natural selection for its intellectual foundation; it defines life in its terms.[1] Along with what seems to be the great majority of biologists, my support for the theory rests to a greater or lesser degree on five central understandings:

- There is evidence of natural selection in the appearance and activities of organisms past and present.
- Taking the place of natural selection, humans are able to produce desired variants by breeding animals and cross-fertilizing plants, as well as in laboratory experiments.

- The occurrence of mutations, random modifications in life's underlying elements or genes (understood today to be nucleotide sequences in the DNA molecule) that serve as the substrate for natural selection.
- Natural selection stands alone, the only mechanism science is able to offer to explain evolution; at present there is no credible alternative.

And finally,

- Selection for survival and reproduction are everyday facts of life known to us all.

But if I am a devotee of a belief common among scientists, why do I feel obliged to protest my fidelity? The reason is that in what follows I identify a serious problem with Darwin's theory—an apparent incompatibility between natural selection and the evolution of the mechanisms of reproduction. Still, why is the protestation necessary? Aren't scientists supposed to be skeptical of their own theories, beliefs, and constructions; to consider chinks in their armor; to identify and question their premises; to confront the evidence that supports them with penetrating, even acid disbelief; and to pursue their exposed weaknesses and inconsistencies with intellectual fervor and the cold, dispassionate eye of reason? Aren't all scientific ideas, however well ensconced, forever tentative, eternally vulnerable to being shown false? Faith in and devotion to an idea is not part of science.

And yet, an avowal is prudent. There are two reasons. One is ordinary, the other extraordinary. As for the ordinary reason, the idea that science is a purely rational enterprise, a pursuit skeptical of its own ideas and evidence is merely what science *should* be. As practiced, scientific belief is neither simply nor directly the product of reason or an objective evaluation of evidence no matter how popular the belief. A great observer of science, Thomas Kuhn, impressed on us what should be obvious. Like all human activities, science is unavoidably a reflection of our "fallen" nature. We practice science as we live life with all our flaws, *even when considering scientific ideas*.

According to Kuhn, science is a series of apprehended beliefs shared by communities of experts that serve as the motive force for study in an area.

He borrowed the term *paradigm* from the social sciences to describe such beliefs. As Kuhn outlines in his classic *Structure of Scientific Revolutions*, his paradigms are based on reason and evidence but also on common assumptions and suppositions, not to mention outright biases and every conceivable political and social consideration.[2]

Perhaps the most remarkable thing about the scientific paradigm is its all-encompassing nature. With evidence or in its absence, whether stated or unstated, the paradigm provides a comprehensive explanation for the phenomena it deals with. Once established by authorities in a specialized field of study, it is held to be true by its adherents as a general matter; that is, with only inconsequential reservations and minor open questions or uncertainties.

Whatever the weaknesses of a particular paradigmatic system, faith in it may be held with unyielding ferocity. Given such an attachment and partiality, questioning the paradigm and the evidence on which it is based is rarely, if ever, greeted by a shared bond of curiosity among scientists. More than likely the response will be defensive. There is often frantic resistance among supporters to it being questioned. According to Kuhn, the inquirer, the one raising the uncomfortable question, is most often an outsider, and if he persists, he becomes an object of ridicule, scorn, and suspicion.

AN EXTRAORDINARY REASON

This "ordinary" reason for my avowal, as consequential and important as it is, is insignificant compared to the extraordinary one. It is the result of the unique impact and impassioned disagreement that Darwin's theory has visited on the greater world outside of science, on our religious beliefs, our politics and social structure, on how we understand others and ourselves. Darwin feared that his ideas would be controversial, especially in regard to religious views of the nature and origin of humankind, and this was at least in part responsible for his decades-long hesitancy to publish them. Though his concern was borne out by subsequent events,[3] it is hard to imagine that he envisioned the enormous, long-lived uproar that has ensued. While the character of the conflict has shifted from time to time, some 150 years later disputes about the theory remain intense and without a sign of resolution among the disputants.

The major cause of the conflict is the theory's wholly mechanical explanation for life's evolution. Its account of evolution does not rely on planning and action by some external agency, conspicuously by God. Evolution simply occurs. It is nothing more than chance and opportunity acting on matter. Natural selection, its instrument, explains how species came into being in an undirected, unmindful fashion, driven solely by physical forces.[4] This presented a great challenge to religious views of humankind's origin in Darwin's time and made religious stories of life's genesis seem little more than fanciful tales by people who were ignorant of science.

Darwinism, as it came to be known, was celebrated as science, in contrast to religious superstition. Many clergy, particularly when the theory first saw the light of day in mid-nineteenth-century Britain, attacked it with passion, condescension, and at times rage. Since then many, though by no means all, Western religious traditions have found ways to incorporate its precepts into their theology. Rejection has led to inclusion.

In addition, for both better and worse, the theory became part and parcel of the political and social turmoil of the time. Amid the Industrial Revolution and the Marxist fervor of mid-nineteenth-century Britain, natural selection and its popular corollary or synonym, "survival of the fittest,"[5] was thought to serve as an intellectual justification for unfettered capitalism; for the abuse of industrial workers; and more generally for the mistreatment or, at least, lack of concern for the less fortunate. Science instructed us that this simply was the way things were and were meant to be. The circumstances of our lives, whether advantaged or disadvantaged, were merely a matter of biology. Some of us were better suited for life than others.

Curiously, given this assessment, Marxists and socialists were also delighted with the theory. While they rejected capitalist arguments for it with unrestrained antagonism, they proudly proclaimed that science had, at long last, refuted the backward-looking ideas of religion that had kept humanity in shackles for so long. Darwin had opened the door to a bright new socialist future. His theory had defeated religion.

In more abstract terms, at one and the same time the theory served as a scientific rationale for the primacy of the individual, for individual freedom (in the survival of the fittest) *and* the bonds of community. If this was not irony enough, socialists and opposing clergymen pointed to the same thing, altruism, to make the case both for and against the theory. To the former it showed that communal action had selective advantage, while to the latter the human traits of charity and love demonstrated that the theory was false. They thought that a mechanism as coarse as natural selection could not have led to such an admirable quality.

Not only was the enormous influence of Darwin's theory on society forged in the furnace of conflict, as said even after all these years, debate and argument continues, somewhat abated but superheated nonetheless. In recent years there has been an attempt by certain public school districts in the United States to introduce the teaching of Genesis into biology classes as a counterweight and alternative to Darwinism, as if religion were science, and in this odd way, odd because religion is a matter of faith and science is not, ensure that students are made aware of the theory's hypothetical nature—it is after all merely a theory, an unproven idea.

A new intellectual movement, or rather a new incarnation of an old one, Natural Theology, emerged that was now dubbed Intelligent Design that planted its feet in both science and religion. It argued that the haphazard nature of natural selection was simply inadequate to explain the evolution of life's complex forms, mechanisms, and processes. A successful explanation requires the intercession of a plan or design. Physics and chemistry are just not enough.[6]

Teachers of science as well as biologists of all stripes have frequently reacted to the attempt to introduce such a discussion into our schools as nothing less than an assault on science that as such needed to be vigorously repelled. The reaction has often been quite emotional; the zeal of its opponents matching that of its proponents, even when the language has been scientific. Rather than seeing the intrusion as a golden opportunity to teach students a valuable lesson about the difference between religion and science, teachers and scientists have at times become strident, even dogmatic, insisting that Darwin's theory has been proven true, that it is a scientifically certified fact of nature that is not to be toyed with.

Regrettably, all too frequently the theory has been defended in much the same way that some religious fundamentalists defend religion, with shrill assertions and intellectual rigidity. This made it seem that the conflict was between two competing belief systems, religion and science. One had to be right, the other wrong. A choice had to be made. This, rather than understanding that the differences were of viewpoint and intention, not right or wrong. As Stephen Jay Gould called them, science and religion were two incommensurable "Magisteria," two overarching philosophical systems with different goals and perspectives.[7]

Whatever your point of view, given these circumstances, for a scientist to raise questions about the theory of evolution is extremely awkward. For those who wish to see it that way, any questioning can be thought of as heresy, as siding with religion against science, even of acting against reason. Which side are you on?! Though undoubtedly very powerful, this dialectic must be rejected; it demands a false and intellectually empty choice.

Given both the ordinary and extraordinary reasons for an avowal of fidelity, it seemed wise to make my devotion to the theory as clear as possible before considering the problem, hence my cry—"I am a Darwinian." My aspiration in what follows is that of science, not religion. My hope is that the paradox can be resolved or if not, at least forthrightly articulated. And further that the effort to do so will enrich our understanding of life's evolution and make Darwin's and Wallace's remarkable insights more fully descriptive of nature's character.

Chapter 3

A LOOK INSIDE DARWIN'S THEORY

Is the Theory of Evolution by Means of Natural Selection Internally Consistent?

No conception can contain anything which contradicts its definitions, i.e., the sum total of its predicates, neither can an existence contain anything which might become a cause of its destruction.

Arthur Schopenhauer, writing about Spinoza's ideas in *On the Fourfold Root of the Principle of Sufficient Reason*, 1813

Scientific theories are judged in two ways. Externally, by observation and experimentation we seek evidence of their goodness of fit to nature, while internally we examine their consistency—are a theory's various aspects consistent with each other, in other words, do its propositions contradict each other? A theory that survives tests of its goodness of fit to nature has a claim on being correct, but the logical consistency of a theory in and of itself allows no such claim. A theory may of course be logically consistent and totally wrong. Conversely, even if a great deal of experimental evidence supports a theory, if it is not internally consistent, it must be misconceived. Evidence, however compelling, just cannot save a theory that is internally contradictory. In this case, its soundness, its validity depend not on evidence, but on its logical makeup.

Of course Darwin's theory is not an exception to this rule. As with all scientific theories, to be useful it must be internally consistent. This has not been perceived to be a problem. In all the years since the theory was first proposed by Darwin, supporters and critics alike have in the main agreed that its various elements and their relationships to each other are harmonious. They present no significant internal contradictions or inconsistencies. The question was not whether they were consistent with each other but whether they were true.

ELEMENTS OF THE THEORY

Before we examine this proposition, we should be sure that we understand the theory's core elements, what it is that must be internally consistent. Though, as we shall see, modern evolutionary genetics offers a somewhat different account, the traditional list, Darwin's and Wallace's list contains five essential features—variation, natural selection, reproduction, heredity, and adaptation.[1] As they understood them, and in the customary sense, they are as follows:

- Variations are differences in the appearance and activities or, in its most general meaning, the properties of organisms.
- In natural selection some organisms, particular variants, are favored over others in a struggle to survive all sorts of challenging environmental (natural) circumstances.
- If they survive and reach sexual maturity, they reproduce. Reproduction is the generation of new organisms from existing ones.
- Reproduction involves the transmission of the parents' features to their offspring, i.e., their features are heritable, even though the offspring are not exact replicas of their parents.

And finally,

- Adaptations are the features of organisms that help them survive the forces of natural selection.

To this we can add:

- Variations arise from mutations, randomly produced structural changes understood today to be modifications in the nucleotide sequence of DNA, as well as from the mixing and matching of the parents' genes during sexual reproduction.
- The circumstances that drive natural selection are the predatory or aggressive behavior of organisms; competition otherwise for limited food supplies; and exigent properties of the physical world that lie beyond living things, such as temperature.
- Because the life of organisms is exceedingly brief compared to the millions to billions of years required for their evolution, continuity is provided by generational renewal or reproduction.

And finally,

- Adaptations are found in why we do what we do.

EVERYDAY FACTS OF LIFE

Significantly, the central features of this enormously important scientific theory are also everyday facts of life that are known to all of us. They are not obscure concepts only accessible to the highly trained expert. As we live life, the existence of biological variation and reproduction are self-evident. As is the fact that some organisms are more successful in surviving life's vicissitudes (natural selection) than others (are better adapted) and hence are more likely to reach maturity and reproduce.

Equally straightforward is the way in which the theory juxtaposes four of its essential features. *Natural selection* could not exist without *variations* to select among, nor could more *adaptively* able organisms emerge in the absence of selection. Selection forms a bridge between variations and adaptations, acting on the first and giving rise to the second.

As for reproduction, as said, life is short-lived. Sooner or later, indeed inescapably our adaptive capabilities prove inadequate and we succumb to the forces of natural selection, that is, we die. As a result, for living things to endure they must be continuously replenished. As selection winnows our number, reproduction renews and expands it. Moreover, it is also self-evident to us that reproduction, at least sexual reproduction, is a critical source of variety. As said, in it, the features of the parents are mixed and sorted to create new individuals with similar but distinctive traits.

Taken together, these commonplace understandings seem quite enough to conclude that Darwin's theory is internally consistent. After all, most of its elements are not abstract notions but experiences of everyday life that as such *must* be compatible with each other. What is more, they appear to depend on each other in just the way that the theory proposes.

THE REST OF THE THEORY

However, Darwin's and Wallace's compelling insight is not found in these commonplace discernments, in these plain facts of life, but in what they inferred from them. Their inference, at once magnificent and extravagant, was that when taken together variation, natural selection, reproduction, heredity, and adaptation are responsible for nothing less than the evolution of life on earth, or as Darwin cautiously called it "descent with modification."[2]

In making this claim, they understood that the mere presence of these elements was not sufficient cause. They would not in themselves result in evolution. For that natural selection had to have direction across time. It had to yield something. There had to be *progress*. Otherwise, even if natural selection engendered effect upon effect, change upon change, adaptation upon adaptation, killing some organisms, allowing others to live, all that would be achieved would be the replacement of one by another. There would be modification, but no descent, no evolution. For that, something had to be accumulated, not merely traded.

According to the theory, what accumulated were *adaptations*. Over countless generations organisms came to have more of them, and they were more effective, resulting in creatures better able to withstand environmental challenge. And so, adaptations evolved and there was progress.

One more thing was needed to complete the theory. For adaptive advantage to accrue, past challenges had to provide some guide to future experience. Otherwise, what was adaptive in the past might not be adaptive now. If that were the case, how could adaptive progress be made? How could evolution occur? With an ever-changing environment, there could only be modification; again, change without descent. This meant that for evolution to take place, natural selection could not be wholly unpredictable or random as is often imagined. Vital and inescapable among its guides were the laws of mathematics, physics, and chemistry. They had to be obeyed. Life's forms and functions had to follow their dictates.[3]

The theory claimed that if all this were true, evolution would occur, and would occur automatically. There was no need for any other agency, for external direction, design or purpose, religious or otherwise. Life would simply evolve. Most importantly, the theory explained evolution in purely physical or mechanical terms as science is constrained to do. Putting aside French naturalist Jean-Baptiste Lamarck's theory of acquired characteristics, the theory of evolution by natural selection is not only science's favored explanation for the materialization of life's many forms and functions; it is its *sole* explanation. This, the theory's uniqueness, perhaps more than anything else accounts for its wide acceptance among scientists.

QUESTIONS ABOUT NATURAL SELECTION

Yet curiously the theory's initial acceptance was not due to recognition of its central premise—that natural selection is the mechanism of evolution—but to the simple realization that biological evolution occurred and did so over extremely long periods of time, millions, or as we now know, billions of years, not the six days allotted by Genesis. In fact, though the occurrence of biological evolution is no longer questioned other than by the most obscurantist critics and the most severe and insistent skeptics, natural selection's role and character, its agency, was and continues to be hotly debated.

To this day it is the subject of intense and impassioned argument and analysis both within and outside the academy. Many questions have been raised over the years. Some of the more durable are the following:

- Are there causes of evolution other than natural selection?
- Are adaptations its sole products?
- How does it produce speciation?
- Does it act in a gradual and relatively continuous fashion, or in fits and starts, or both?
- Can natural selection account for gaps in the historical record?
- Can the many seemingly haphazard events of natural selection account for systems of the exceptional organization, complexity, and even perfection found in biology?
- Can natural selection explain altruism, behavior that expresses "unselfish concern for the welfare of others"?
- Does natural selection act on parts or, conversely, groups of organisms, in addition to individuals?

Finally, and perhaps most pointedly,

- Is natural selection merely a metaphor without substance in the material world or, worse yet, a tautology?

Many of these questions were posed by Darwin and Wallace themselves and have shaped debate about the theory since. I will expand on some of them in the next chapter, but whatever the answer to a particular question, whether we think it affirms the theory or proves it false, these questions as well as others that I might have included are about the theory's goodness of fit to the *observable or external world*. None concern its internal consistency. This is, as said, because supporters and critics alike have believed the theory to be internally consistent.

A CURIOUS DISSONANCE

Yet there is a curious and rather obvious dissonance. Though both natural selection and reproduction are needed for evolution to take place, strikingly and self-evidently they are events of very different kinds:

- First and foremost, natural selection concerns the survival of an existing organism, whereas reproduction is about the production of a new one.
- Likewise, the mechanisms wrought by natural selection seek to ensure survival of an existing individual, whereas those of reproduction concern the production of new ones.
- Natural selection acts on individuals, even as it affects groups, while reproduction, though carried out by individuals, is about groups, family lines, and species.
- Natural selection acts in the moment, whereas reproduction secures the future.
- The object of natural selection exists when it acts, whereas reproduction is about objects prior to their realization. Indeed, in sexual attraction, it is about the yet to be conceived, the nonexistent, the only imagined, even the unimagined.
- The traits and mechanisms of the nonreproductive or somatic features of life, the products of natural selection, are the means whereby organisms survive, whereas reproductive traits and mechanisms are the means that allow for the production of new generations.
- Natural selection produces progress in a negative and harsh fashion, eventually destroying all living creatures in the process, whereas reproduction is resolutely positive because, whatever its other flaws, it creates life.

Finally, and most importantly,

- Natural selection produces its results without intention, while by all accounts reproduction intends to produce offspring.

Natural selection and reproduction are not only stunningly different; they are contradictory and in some respects opposites. Critically, whereas reproduction seems to have a goal, production of the next generation and the continuance of life, natural selection cares not a whit about the future or for that matter the survival of life, it merely acts on what is before it. This lack of intention is not incidental but is central to the theory. It is the key element in the proposition that natural selection impels evolution. If selection was goal driven, there would have to be a "setter of goals," God or some godless designer. And without doubt the most important thing about Darwin's theory is its exclusion of design or purpose. Evolution just occurs.

Given that the two occurrences are so contradictory, it is appropriate to ask how natural selection can have produced the reproductive features of life. We will ask and attempt to answer this question in the upcoming chapters, but before we do, we need to examine what we mean by natural selection in a little more detail. That is the subject of the next chapter.

Chapter 4

THE MEANING OF NATURAL SELECTION

Defining Natural Selection

> *The preservation of favourable variations and the rejection of injurious ones, I call Natural Selection.*
>
> Charles Darwin, *On the Origin of Species*, 1859

Along with *cell* and *gene*, *natural selection* is one of the three fundamental concepts of biology. It is thought to explain that great wonder of nature . . . the evolution of life. The idea is highly attuned to our modern scientific sensibility in that natural selection is understood to account for biological evolution in terms of disinterested material or physical forces. Despite its importance, the meaning of the term has been a matter of substantial confusion and not a little controversy since Darwin first used it.[1] What follows is a brief guide to the most common usages.

AN ANALOGY AND A METAPHOR

In proposing natural selection as the agency of evolution, Darwin drew on an analogy and a metaphor. The analogy was geological and concerned the effect of physical forces on matter. In the world of inanimate objects, natural selection is the kinetic energy of things acting differentially or disproportionately on other things. It is rain or flowing water eroding rocks made of granite or clay. It is a moving mass causing others to move, while leaving others seemingly unmoved. It is the temperature of the atmosphere vaporizing water while at the same time solidifying the planet's core substances as they emerge from volcanoes.

The metaphor was the breeding of domestic animals. In this case, people play the role of, if not of God, then nature, by selecting traits of breeding pairs to meet human needs. Darwin described this pursuit and compared it to what occurs naturally as follows:

> Can it, then, be thought improbable, seeing that variations useful to man have undoubtedly occurred, that other variations useful in some way to each being in the great and complex battle for life, should sometimes occur in the course of thousands of generations?[2]

In other words, while humanity acts to bring variations into being to suit its requirements, nature produces variations helpful in the struggle for survival. The same metaphor applies to horticulture, where plants are modified to suit our desires and needs.

DEFINING NATURAL SELECTION

Even if natural selection were a scientific idea of no special note, we would still need to be as clear as possible about what we mean by the term to have any hope of evaluating its soundness and applicability. The epigraph at the head of this chapter is Darwin's sparse initial definition in *On the Origin of Species*, echoed in the last chapter as one item on a list of the elements of evolution comprised additionally of variation, reproduction, heredity, and adaptation: "In natural selection some organisms, particular variants, are favored over others in a struggle to survive all sorts of challenging environmental (natural) circumstances (see p. *XXXXXXXXX*.)"

This was not Darwin's only definition. He described three different though related occurrences that he called natural selection. In the first and most commonly understood, organisms of different species engage in a competition for food (being or getting) or otherwise seek to survive their environmental circumstances. A gazelle running away from an attacking cheetah is a familiar example. The outcome of their engagement represents natural selection. Either the cheetah catches the gazelle and kills it, or it is too slow and eventually dies of starvation. If enough life-ending events of this sort take place for a species, it becomes extinct and is no longer part of the long march of evolution.

The second definition involves selection among individuals of the same species or, more narrowly, the same reproducing group. As the cheetah attacks the herd and catches a slow gazelle, often a young or old one, faster gazelles speed away, escape, and survive the attack. In this, nature has made another selection, a selection within the group of gazelles. Everything else being equal, the speedier, more adroit gazelle survives, while the slower, less adroit gazelle, having been killed, obviously does not. It has been selected against.

Thus natural selection has both interspecific and intraspecific manifestations. They are mutually inclusive and concurrent. Indeed, they are different aspects of the same event. As the gazelle competes with the cheetah, it also competes with other gazelles. Natural selection judges it both ways. But either way, selection concerns survival. The gazelle survives or it does not. It does not matter whether we think of it in relation to cheetahs or other gazelles. Simply, there are winners and losers.

THE THIRD WAY

Despite the popular belief that thereby natural selection is the *cause* of evolution, it is not. We can say that it is its *agency*, but not its cause. This is because what natural selection does, all it does, all it can do is choose. It neither creates nor transforms, and both creation and transformation are necessary for evolution. What is missing is reproduction. Without procreation, there is no evolution, and the effect of natural selection's choices, whatever they are and however hard won, would end with the death of the existing generation. Evolution requires both. They act in sequence. Either first there is reproduction and then natural selection, or first there is natural selection and then reproduction. For the latter, first the victors are selected, and then reproduction ensures that their species or group continues. Though selection takes place among individuals, it is the reproducing group or species that evolves.

This brings us to Darwin's third and most perplexing version of natural selection. In it reproduction and natural selection are seen as two aspects of the same process. Since reproduction only takes place within a species, this version limits selection to a choice among its members, more specifically to particular subgroups. In this view, it is heritable differences in the traits of subgroups that serve as the basis for natural selection.

Initially the differences between the groups may be small, perhaps barely noticeable, but are nonetheless sufficient to provide one group with some sort of advantage over others. In the fullness of time, over many generations, these differences become more substantial, more clear-cut and well-defined. Eventually they give rise to new species, to speciation, and thereby to biological evolution.

This is thought to come to pass in two ways. In the first, a particular, new, or newly emergent form becomes dominant and replaces those less able to survive and reproduce. Alternatively, the new form separates, rather than dominates. It may find a new habitat or new niche. This produces a geographical isolation that limits reproduction to other members of the same subgroup. Separation can also be social, with partners being chosen on the basis of the desirability of divergent physical or behavioral characteristics; or temporal, with different individuals seeking sex at different times, different times of year or even different times of day. In whatever way, as the generations pass, differences between the groups increase, until they are sufficiently unlike each other, as well the ancestral type, to become a distinct species. Whether this occurs by displacement or isolation, new species arise as the result of reproduction.

SELECTION BY GROUP

Understood in this way, natural selection is not only about the fate of cheetahs and gazelles as they live life but also about future generations of cheetahs and gazelles. Not only may their kind continue or become extinct; they can evolve into new species. This version of natural selection is not just about the survival of existing individuals, about their personal struggle in the moment or across time but is also about the survival of biological groups over generations. Ultimately, competition for survival occurs between *groups* within a species, not among individual organisms. Natural selection is what happens to the *group*, and the path of evolution is its path. It is selected for or against, whatever the fate of individuals.

From this perspective, despite their material independence and the fact that they occur in sequence, reproduction and natural selection are aspects of the same phenomenon. Their purposes are entangled. And if natural selection is about the fate of groups as well as individuals, then it is also unavoidably about reproduction. Similarly or conversely, reproduction is not just about producing new creatures; it is an element of natural selection.

And so, unlike Lamarck's theory of use and disuse, the theory of acquired characteristics, which envisions organisms gaining useful features and losing those that are not useful during their lifetime, in the Darwinian concept of evolution it is not the individual that evolves (it dies as is) but rather the group or species to which it belongs. Evolution is a property of the group, and the group can only evolve as the result of reproduction. This is all well and good, and is well known, but how does it work? What is the mechanism of group selection?

If groups are aggregates of individual organisms, what happens to the group is based and based ineluctably and solely on what happens to its individual members. In other words, though natural selection affects the fate of the group and it is the group that evolves; it does *not* act on it. The individual and it alone is the object of its ministrations. The individual is the unit of selection, and reproduction has no direct role in that. While reproduction is required for evolution, the selection of traits is the consequence of natural selection's action on individual organisms, not groups.

This appears to be how Darwin imagined group selection—the effect on the group of what happens to individuals. He did not propose, as others have since, that there is an actual selection among groups as entities unto themselves (much more about this just below and when we get to reproduction).[3] From Darwin's viewpoint, natural selection and reproduction retained their separate and independent character, each following the other time after time. Over the course of the ages their actions yielded evolution.

THE MODERN SYNTHESIS

In a different conceptualization of the third way, natural selection and reproduction are not merely procedurally connected; they are also materially merged, fused in fact. This perspective arises out of the great accomplishment of R. A. Fisher, J. B. S. Haldane, and Sewall Wright in the 1930s reconciling Darwin's view of evolution by means of natural selection and Gregor Mendel's genetic view of inheritance.

If evolution is driven by natural selection and inheritance is the result of reproduction, then Mendel's and Darwin's theories offered conflicting explanations for life's genesis, or so it seemed. Independently, Fisher, Haldane, and Wright came to the conclusion that this clash was only apparent. It disappeared once one realized that natural selection and inheritance were different aspects of the same phenomenon. Evolution was the result of a genetically (reproductively) based natural selection.[4]

Known as the modern synthesis, this insight moved natural selection's locus of action from the expressed features of individuals, from life's phenotype, where Darwin had firmly placed it, to its genes or genotype, to the underlying source of these features.[5] By the 1950s genes were known to be particular nucleotide sequences of the enormous DNA molecule. It was DNA that coupled natural selection and reproduction. DNA's gene sequences singly and together (the genome) and changes in them (mutations) gave rise to the features of the organism on which natural selection acts, while duplication of the same DNA was the basis for reproduction. Natural selection and reproduction were joined in the properties of the DNA molecule.

Life was carried forward through DNA. It was DNA, not organisms, that survived and evolved. Organisms were merely containers for the evolving DNA. As evolutionary biologist Richard Dawkins explained:

> We are machines built by DNA whose purpose is to make more copies of the same DNA. . . . This is exactly what we are for. We are machines for propagating DNA, and the propagation of DNA is a self-sustaining process. It is every living object's sole reason for living.[6]

This gene-centric view of evolution can be understood in three ways. In Darwinian terms, while natural selection may act on the expressed features of organisms, it is really the underlying genes that evolve. Alternatively, natural selection acts directly on the genes, as well as on their expressed features, selecting among the DNA of different organisms. Finally and most radically, the inherited information of the organism embodied in its DNA evolved in its own right. Darwin's natural selection and the phenotype or expressed features of organisms were not involved.

Though these concepts are straightforward, how the proposed events would take place is not. For instance, how can underlying genes (DNA) be the objects of selection if the environment acts not on them but on the expressed features of the organism? Or how would natural selection select among DNA molecules in their own right other than in terms of their chemical properties, such as their stability? And finally, without natural selection, how would evolution be achieved, by means of what alternative agency of selection?[7]

THE VIEW FROM EVOLUTIONARY GENETICS

Some evolutionary biologists have tried to remove natural selection from evolutionary theory,[8] replacing it with three principles. Evolutionary biologist Richard Lewontin gave this easy-to-understand account in a *New York Review of Books* article. According to Lewontin, "The modern skeletal formulation of evolution by natural selection . . . stripped of its metaphorical elements" is comprised of three principles:

1. The principle of variation: among individuals in a population there is variation in form, physiology and behavior.
2. The principle of heredity: offspring resemble their parents more than they resemble unrelated individuals.
3. The principle of differential reproduction: in a given environment, some forms are more likely to survive and produce more offspring than other forms.[9]

As an addendum, he includes a fourth principle—the principle of mutation—the introduction of a new heritable variation. Whether there are three principles or four, and even though they are offered as an expression of the theory of "evolution by natural selection," natural selection is not among them. The first and second principles tell us that variation and heredity are critical to evolution. And that is certainly true. But neither alone nor together, even if we include mutations (and the sorting of genes during reproduction), do they give rise to evolution. Though hereditable variations endure by means of reproduction, endurance is not evolution. All reproduction does is make more of what already exists, and whatever changes occur during the process (genetic rearrangements) or prior to it (mutations) are not directed but random, the result of chance, and that, too, is not evolution.

It is in the third principle, "differential reproduction," that we find evolution. In this view reproduction is more than the mere creation of new individuals. It is termed "differential" because it determines the evolutionary path, with some reproductive groups being favored over others. The choice is found in the number of offspring they produce. The characteristics of the most fecund group reflect evolutionary progress.

But how can this be correct? After all, the number of offspring varies enormously between species, from one to many hundreds, and these differences in fecundity do *not* reflect differences in the evolutionary worthiness of a species. Species that produce few offspring are not de facto less fit, less likely to survive (or more likely to become extinct), than those that produce many.

But this is not what Lewontin means by differential reproduction. He is not referring to these differences but to procreative differences *within*, not between, species. It is in exceptions to the rough identity in the number of offspring produced by a given species and its subsumed reproductive groups that we find evolutionary selection by means of differential reproduction. The more reproductively successful subgroups flourish and embody evolution's forward march.

In this it is not the fecundity of the individual organism that drives evolution but rather the number of progeny that different intraspecific reproductive *groups* produce. Differential reproduction is really about group selection. The presence of the more productive group is differentially enhanced relative to those less endowed, and with time it comes to dominate the population.

But success is not merely a matter of producing more progeny. Reproduction is not differential simply because the physiology or practices of one group happen to yield more offspring. This number does not itself determine the group's success as it faces the future. There is another player and it is critical. It is found in the events that bracket reproduction, that *precede and follow* it. And of course what precedes and follows reproduction is nothing less than the life that is lived. However many progeny are initially produced, the life lived determines how many members of a particular group survive the journey to sexual maturity and are able to reproduce. And what determines this, what determines survival is none other than natural selection! It is its action on individual organisms that controls how many members of the group are still alive when it comes time to reproduce.

So Lewontin's third principle is not merely about reproduction. As he says, differential reproduction is about which "forms are more likely to *survive* and produce more offspring than other forms" (italics added). The word *survive*, though delicately introduced, is the key to the evolutionary power of differential reproduction. If more members of a group successfully negotiate life's vicissitudes, then more will be available to engage in sex, and consequently the group will be more productive, and of course vice versa. And the survival of extant organisms is the business of natural selection.

It turns out that replacing the term *natural selection* with *differential reproduction* is either semantic or the consequence of a desire to place the emphasis on reproduction rather than on natural selection. Differential reproduction is what comes after natural selection has had its way. In itself it is neither an evolutionary process nor a means of selection. It is the means by which the action of natural selection on *individuals* affects the *group*. Natural selection determines evolution's direction, while reproduction ensures its occurrence. I will return to this idea later when we consider the evolution of life's reproductive mechanisms in section V, "Purpose and Reproduction."

In any event to talk of differential reproduction rather than natural selection is seductive for a reductive biology. It makes evolution seem more like an automatic unfolding of the mechanisms of reproduction and genetics, rather than being dependent on natural selection's external uncertainties. Natural selection seems like weak tea, too vague and unconnected to the mechanical world to determine the path of evolution. Certainly, variation and heredity, firmly embedded in the reductionist view as mutations and genes in the DNA molecule, are critical. And the incarnate mechanisms of genetics and reproduction seem a better way to understand how things evolved than natural selection, so inadequately (and uncomfortably) defined as the ability to survive.

But of course, as said, survival is necessary for reproduction. Just as there can be no evolution without reproduction, reproduction cannot occur without living things to carry it out. And of course they can only do so if they survive life's travails to reach sexual maturity. However messy, natural selection cannot be excoriated, it cannot be done away with. This is why the three principles are announced as an expression of the theory of "evolution by natural selection." Even if natural selection is not one of the theory's principles, it is not only needed; it determines evolution's path.[10]

THE MECHANISM OF NATURAL SELECTION

Natural selection presents another discomforting oddity. While perhaps more than anything else the raison d'être of Darwin's theory of evolution was that natural selection provides a mechanical explanation for biological evolution, natural selection has no material incarnation, no physical basis. Like mutations and genes, it is a reductionist idea, but unlike them it has no embodiment. It is not a mechanism at all, but a principle. Though broad and all-encompassing, it is not causative because it lacks a causative instrument. Unlike a mechanism, natural selection cannot be reduced to something else. All we can say is that the principle of natural selection is the principle of natural selection. It is nature's mindless *choice* about the fate of things in the material world, and that world includes living things.

THE ORIGINAL DARWINIAN SIN

And so for all its choices, natural selection is *not* the cause of evolution. It is evolution's agency, it sets its path, and as such is necessary for its occurrence, but it is not its cause. For that we need reproduction and its genetics. This brings us to the crux of the confusion about the term *natural selection*. It can be traced to Darwin's "original sin." As pointed out, Darwin's proposal of natural selection as evolution's agency was based on an analogy and a metaphor. The analogy was to events in the inanimate world, to a geological natural selection, in particular to the differential effects of forces on objects. Though the forces were common, the responses to them were disparate, and hence nature-made choices. The metaphor was that natural selection was like humans husbanding animals, in which case we played the role of natural selection in producing change.

In the analogy, natural selection simply means that nature selects among different material objects and that this choice proscribes their fate. It is nothing more than a fated, but unintended, choice. It is a "natural" selection. On the other hand in the metaphor not only do husbanding humans make choices, their choices are followed by reproduction among the pairs they have chosen. In this case, selection includes reproduction. Consequently, it is not just about a choice but is also about reproduction and hence evolution. Indeed, reproduction is the cause of evolution. If reproductive mechanisms are included, the principle of natural selection not only sets evolution's direction; it serves as its material or mechanical basis.

Darwin was not ignorant of this distinction, but nonetheless he used the same term with its two different meanings interchangeably. And so did those who followed, including the current author. It means nothing more than selection here, but there it means evolution, that is, selection with reproduction. This did not really matter because whenever the term was used in reference to biological evolution, reproduction was implied whether or not it was expressly noted. With this discussion of natural selection in hand, we are now prepared to tackle the core subject of *The Paradox of Evolution*, the evolution of the mechanisms of reproduction and the role of "natural selection" in it.

SECTION II

REPRODUCTION

Chapter 5

ONE FOR ALL

Did Natural Selection Produce the Mechanisms of Reproduction?

Inasmuch as peculiarities often appear under domestication in one sex and become hereditarily attached to that sex, the same fact probably occurs under nature, and if so, natural selection will be able to modify one sex in its functional relations to the other sex, or in relation to wholly different habits of life, as is sometimes the case with insects.

Charles Darwin, *On the Origin of Species*, 1859

However obvious the differences between natural selection and reproduction may be, as said, it seems to be the common, if not the universal, view among scientists that life's reproductive features, those features responsible for producing offspring, are the product of the same agency that is responsible for the creation of its somatic or nonreproductive features. That is to say, if we know anything about evolution, it is that natural selection fashioned the structures, mechanisms, and processes of reproduction just as it did all others. Yet beyond statements like Darwin's in the epigram above, as best I can tell the basis for this belief has never been articulated nor has the presumption been subject to serious critical examination.[1] Still, the reasons are obvious and indeed persuasive.

FALSE CHOICES, PARSIMONY, AND CONSANGUINITY

To begin, we can say that the comparison between natural selection and reproduction laid out in chapter 3, however undeniable, is beside the point. It compares apples to oranges; it confuses what something does—reproduction—with what brings it into being—natural selection. What brings something into being need not have any meaningful connection to what it does. For example, knowing that the heart is the result (not the product) of natural selection has no express bearing on the fact that it pumps blood. There is no reason to suspect that things are different for reproduction. There is no reason for thinking that natural selection has any bearing on the character of the evolved traits.

Nor does the fact that natural selection does not care about the future while reproduction is all about it necessarily distinguish one from the other. According to Darwinian theory, despite natural selection's indifference and detachment, it charted life's evolution. However unintentional, natural selection is about the *future*. Why can't the same indifference apply to reproduction? Why can't reproduction, like everything else about evolution, be the consequence of purposeless natural selection despite the fact that it is responsible for ensuring life's future? Reproduction's seeming intentionality may simply be a misapprehension.

Then there is the argument from parsimony. In the fourteenth century Sir William of Ockham said in his famous rule, or razor, that science must hew to the simplest explanation for natural phenomena. Or more formally, that everything else being equal, to warrant its name, science must favor the theory that requires the fewest elements or propositions. Certainly, a single agency being responsible for the evolution of both the reproductive and nonreproductive features of life is more parsimonious than different agencies being responsible for each. As a matter of parsimony, and without specific evidence to the contrary, science is obliged to conclude that whatever produces the nonreproductive features of life also produces its reproductive ones.

And surely any claim that the evolution of reproductive features is the product of some other agency should be accompanied by some notion, however preliminary, of what that agency might be. Putting aside Jean-Baptiste Lamarck's discarded theory of use and disuse, science has nothing to offer other than natural selection. To our knowledge it provides the only serious causal or scientific explanation for evolution, and that necessarily includes its reproductive features. If this is so, then on what basis do we question its role?

If this is still not sufficient, there is consanguinity, the fact that the physical and chemical nature of the body's reproductive features are of the same sort and arise from the same source as its somatic features. Cells that concern reproduction are in great part comprised of the same substances (e.g., DNA, RNA, proteins, carbohydrates, fats) and structures (e.g., nuclei, mitochondria, Golgi apparatus, endoplasmic reticulum, the cell membrane, fibers and tubules of various sorts) that in great part carry out the same critical acts (e.g., DNA duplication, protein production, metabolism, membrane transport, and cell division) as cells that are not involved in reproduction.

Although special reproductive cells, such as sperm and ova, differ from somatic cells, say, from hepatocytes and muscle cells in conspicuous ways, nonetheless they share the same *foundational* chemical, genetic, and anatomical attributes and arise from the same source, the fertilized ovum. There is no basis for claiming that they are the products of different evolutionary agencies and that as a consequence they are *qualitatively*, that is, *fundamentally* different. The differences that do exist, for instance, between the cells of the testes and the stomach, are no weightier, of no greater significance than those between different nonreproductive cells, for example, between neurons and gastric cells.

And then there is the theory itself. Darwin's theory not only proposes that natural selection applies to reproductive processes; it requires it. Its claim is that natural selection accounts for the evolution of *all* the essential properties of life, including reproduction. It does not exempt it. Consequently, to question the role of natural selection in the evolution of life's reproductive features is to question its role in the evolution of all of life's features. It is to cast collective doubt on Darwin's theory. If, conversely, we accept the theory and the evidence that supports it for the evolution of life's somatic features, we are obliged, at least as a matter of theory, to accept that natural selection played the same role in shaping its reproductive traits.

THE DIFFICULTY REMAINS

However convincing these reasons may seem, even when taken together, they are unsatisfactory. This is not because they are not true but because they are insufficient. None, however valid, actually compares natural selection as the single common cause of life's evolution, what it does, how it acts, to the products of reproductive evolution. The question "How specifically did natural selection produce reproductive evolution?" has not been asked or answered. Do the two make sense as agency and consequence?

The relevant question is not whether a comparison between natural selection and reproduction is apt. It is not. They are clearly events of a different sort. Nor is the rule of parsimony open to question. No doubt one cause is better than two. Nor can we doubt the deep similarities in the material nature of life's reproductive and somatic features and their common source.

The relevant question concerns none of this. It is whether natural selection and reproduction fit together causally, as agency and product, that natural selection is the agency responsible for the various evolved mechanisms of reproduction. To answer it, we have to look at their relationship as it is found in nature. How does one lead to the other, how does natural selection produce life's reproductive features? In seeking an affirmative answer to this question, we must start at the beginning with reproduction's origin and its relationship to natural selection.

INTO THE PRIMEVAL CALDRON

Even here in its origins the idea that reproduction is the product of natural selection faces an overwhelming difficulty—a problem of sequence. Although natural selection existed before reproduction, indeed before the dawn of life, and as mentioned explains all sorts of occurrences in the world of inanimate objects, such as soil erosion and deposition, its role in life's evolution depends, and depends absolutely, on the *prior* existence of reproduction.

As explained, if life had somehow come into being in the absence of reproduction, natural selection's choices would have been fruitless. All life would end, unavoidably, inexorably with the death of the first generation. Our continued existence as living things, our perseverance in the struggle for survival only has evolutionary meaning through reproduction. And although natural selection is evolution's guiding light, reproduction is natural selection's preface. It makes its role in evolution possible.

That is to say, natural selection is only evolution's motive force in the presence of reproduction. Accordingly reproduction cannot have been the consequence of natural selection. Self-evidently an action that depends on the prior existence of something else cannot cause it. For that, reproduction would have to have existed before reproduction came into being, and that is, of course, impossible.

THE ORIGIN OF REPRODUCTION

How then did reproduction arise? What led to its materialization? Though the actual facts are unknown and may be unknowable, it is believed that reproduction began as a chance chemical occurrence. Some four billion years ago a molecule with the ability to duplicate or reproduce itself came into being. The odds of this were remote. As a rule chemical reactions transform matter; they change one substance into another. Substances do not in the normal course of events duplicate themselves. Indeed, we know of no natural chemical compound that has this capability, and this includes DNA.

The common belief is that such an occurrence is so rare that it only happened once, or if not once, then only at one unique time in the history of the earth. It was a singular event. This may seem like a pretty feeble and insubstantial foundation for developing thoughts about the origin of reproduction, an event that is not only rare but that occurred only once. The claim may seem preposterous, especially for something so important. Yet preposterous or not, it is not frivolous. The simple fact is that we know of no other way that reproduction could have come into being (placing it on another planet or in space, as some have, only changes the location of the event, not its character). However unlikely, it seems that this is what must have happened.

If this is not enough, there is another problem. To those who do not know the intricacies of DNA chemistry, it may seem odd to say that DNA does not copy itself. After all isn't that what it is supposed to do? But DNA is only duplicated with the help of special enzymes called DNA polymerases. They are responsible for arranging the four different nucleotide subunits that comprise the DNA chain in the precise sequence specified by the parent molecule. DNA being copied, and copied accurately, depends on these enzymes and the chemical reactions that they facilitate or catalyze.

The difficulty is that DNA polymerases are proteins, and protein structure is specified by the genetic code in DNA; that is, the presence of polymerase enzymes depends on the prior existence of DNA and its code, while the production of DNA requires the prior existence of polymerases. There is a chicken-and-egg problem. In fact, you could say it is evolution's original chicken-and-egg problem. Which came first, the DNA chicken or its polymerase eggs? The explanation usually given is that some antecedent molecule, not DNA (perhaps RNA), had the *intrinsic* ability to copy itself. Over time, the argument continues, this ability was lost. Through however many intermediate steps, DNA, now lacking this ability, replaced the original self-replicating molecule.[2]

NATURAL SELECTION IN THE WORLD OF REPRODUCTION

The advent of duplication did not automatically signal evolution. In a Garden of Eden of plentiful resources and capacious space, where lions, if they existed, could lie down with lambs, there would be no natural selection, and without it there would be no evolution, however "fecund" the reproducing molecules. The replicating objects would simply increase in number, unencumbered. If there were different forms, they would accumulate concurrently and coexist peaceably. All would be equally successful (a happy socialist thought).

Such a halcyon time, if it ever existed, was before competition for survival reared its ugly head, before Darwin's natural selection, before biological evolution, and hence before life and its adaptations came into being. For natural selection to take place competition was needed. That required two things, variety and scarcity—variety in the types of replicating systems to be chosen among and scarcity in what they needed to reproduce. At some point, this came to pass, and life came into being and subsequently evolved.

Variety was introduced by random changes in the chemical and physical properties of the reproducing objects produced by environmental forces, notably by cosmic radiation. This is what we call "mutations." Scarcity had two elements—a dearth of the precursor substances needed for duplication and limited space for the reproducing objects to occupy. As for the former, some systems were able to incorporate scarce molecules more avidly or more rapidly than others. We might say that they behaved "gluttonously," thereby "starving" their competitors. As for space, occupying it was a function of the rate of reproduction. If space were limited to particular ecological niches, sooner or later the more rapidly reproducing objects would fill it, displacing those that reproduced more slowly.

Only with scarcity did the great shadow of life—competition for survival—appear. Reproducing objects could no longer be effortlessly generated. Critical advantage accrued to systems that had the ability to scavenge an increasingly dilute environment for needed precursors (or the ability to modify other substances to form them, as in photosynthesis).

Life and its natural selection were born of the need for precursor substances. Eventually, with continued reproduction, most of the essential precursors were consigned to the replicating objects themselves. When this came to pass, another ability became critical, indeed unavoidable if life was to be sustained. It was the ability to obtain needed precursors from *other* reproducing systems. This meant that it was necessary to destroy others or secure their substance in less fierce ways (as in parasitism). In one way or another, to survive it was necessary to break other organisms down, to digest them. Systems capable of doing this were the first "predators" (from the Latin to "take booty" or "seize as plunder"). Opposing predation, some reproducing systems came to display a defensive capability. They could prevent or otherwise mitigate the destructive effects of the predators and of the environment otherwise.

With these opposing capabilities—predation and defense—biological evolution began. Most of what ultimately evolved, most of the adaptations that we see today, with their many incarnations, mechanisms, and processes, function in the service of these two primal ends—obtaining the substances (nutrients) needed for replenishment, growth, and reproduction, and preventing oneself from becoming them, preventing oneself from becoming fodder for others.

The end result of all this are the somatic properties of life, our nonreproductive adaptations. But however complex, varied, and well developed our somatic adaptations have become, they are merely the handmaidens of reproduction. As explained, survival is necessary for evolution, but it is not sufficient. What it does is make reproduction and hence life's continuation possible. Looked at in this way extant life serves future life and reproduction's overarching transcendent purpose. I will have more to say about this later.

Though it is self-evident that existing life is of value to future life, since without it, there would be no future life, what are we to make of the converse? Of what value is future life to existing life? What is the survival value of reproduction to existing organisms? If natural selection were responsible for the enormous variety and complexity of the mechanisms of reproduction that subsequently evolved, it had to have improved the circumstances of those who engaged in the process, ergo the parents. Were their prospects, was their likelihood of survival improved by their engagement in reproduction, and if so, how? Any claim that natural selection is responsible for the evolution of the reproductive features of life must provide a positive answer to this question.

Chapter 6

THE OTHER BEGINNING

The Reproduction of Cells

Natural selection, on the principle of qualities being inherited at corresponding ages, can modify the egg, seed, or young, as easily as the adult.

Charles Darwin, *On the Origin of Species*, 1859

As with molecular replication, secure knowledge of the earliest mechanisms of cellular reproduction are lost forever in the mist of time. We cannot even be sure that duplicating chemicals antedated the dawn of the cell and cellular reproduction.[1] But whether they did or did not, sooner or later some self-replicating molecules were constrained to the enclosing space of the cell with their fate tethered to *its* process of reproduction.

Two forms of cellular reproduction were possible. One was termed asexual and the other sexual. Sexual reproduction came to be the central reproductive process of complex plants and animals, as well as many single-celled organisms. As anyone who has taken an introductory course in biology knows, in asexual reproduction each cell divides into two daughter cells. Whereas in sexual reproduction two cells, the parents or, in multicellular organisms, special cells derived from each parent such as eggs and sperm fuse with each other and give rise to their mutual offspring. For multicellular organisms this fusion is followed by development, a process of cell division that generates the offspring. During development the replicating molecules (DNA) contained within the cell are dependently sorted, split (evenly) between the cells as they divide.

CREATIVE DESTRUCTION

To determine whether these mechanisms evolved as the result of natural selection, we have to look, as has been said, to the *parents*, not the progeny. The reasons for this are self-evident. First, natural selection acts only on existing life. It has no effect—and how could it—on life that has yet to come into being; that is, on the eventual products of reproduction, the progeny. Second, and equally obvious, it is the reproductive traits of the parents that give rise to the new generation, and that as such would be the objects of natural selection.

In other words, to establish natural selection as the agency responsible for the evolution of the reproductive features of life, it must be shown to provide an adaptive advantage to the parents (or be associated with a property that does). In one way or another reproductive traits must improve *their* chances of survival. As it happens this is *not* the case for any facet of cellular reproduction or, for that matter, for cellular reproduction in its entirety, and the grounds for this conclusion are easy to appreciate.

As new cells are produced, the progenitor or parent cell is *destroyed*. In asexual reproduction it is cut up, and in sexual reproduction it is made a part of something else (the fused cell). Either way the original object no longer exists; it is gone. And naturally a process that destroys something cannot be of value to it. Even if the new entity is initially comprised entirely of the contents of the progenitor(s), as is the case in fusion, having ended its existence, it cannot serve its interests. It cannot provide an adaptive advantage to what it does away with. And providing such an advantage is not merely a paramount feature of natural selection; it defines it.

THE EMERGENCE OF CELLULAR REPRODUCTION

If the mechanisms of cellular reproduction did not emerge as the result of natural selection, then how were they produced, by what physical or chemical mechanism? We do not have to imagine a rare occurrence for this as with molecular replication; there is a very common one—bubbles. If bubbles become too large, that is, if they have too little surface area for their volume, they become unstable and break up into more stable smaller objects. That is, they *divide*. Or if they are too small and in danger of breaking apart because their surface area is *too large* for the enclosed volume, they *fuse* with adjacent bubbles to make a larger, more stable object. Perhaps the original cells were some kind of bubble.

Though this might account for the origin of cellular reproduction, it cannot of course explain the complex mechanisms that subsequently evolved, such as mitosis and meiosis, cell division and cell fusion (fertilization). In them, a variety of specific structures and molecules work in concert to realize a uniquely biological goal. Most consequentially, commonplace physical or chemical occurrences, such as bubbles or chemical reactions, cannot explain their remarkable, almost faultless non-stochastic or nonrandom precision.

What ordinary chemical reaction produces only a single copy of a substance as in DNA replication, and what purely physical process yields objects that always divide in half, but not significantly more or less than half, no less into many pieces? What ordinary stochastic or random physicochemical event would uniformly distribute one copy of DNA to each daughter cell, never more or less? And finally, in what simple physicochemical system would cells only fuse with one other cell, and only very rarely more than one? Such specificity is hard to imagine in terms of the random physical and chemical properties of the inanimate world. Some other explanation is required, and it cannot be natural selection.

THE OBJECT OF NATURAL SELECTION'S DESIRE

This puts us in an awkward spot. Given that an ordinary physical or chemical process is not possible, if we reject natural selection as the agency responsible for the evolution of the mechanisms of cellular reproduction, we are left with only two possibilities. Either there must be some other creative, that is, selective force, or we have to reconsider our rejection of natural selection. We will find ourselves in this predicament time and again in what follows, needing another agency of selection or having to reconsider the rejection of natural selection. As a matter of inclination on each occasion I will try to accommodate reproductive evolution to natural selection before looking elsewhere. Though this is not ontologically required, given the special place of natural selection in biology, the concept certainly deserves the benefit of doubt before being set aside. There are three ways that natural selection can have accounted for the features of cellular reproduction, and, for that matter, for the reproductive features of life in general.

The first is that despite what I have said, the mechanisms of cellular reproduction came into being for purposes that are related to survival, not reproduction. Reproduction would be an afterthought. This is similar to Alfred Russel Wallace's critique of Darwin's theory of sexual selection (see section III, "Sexual Selection"). Though, as we shall see, such sexually infused traits exist and indeed are very common, the central mechanisms of cellular reproduction, the aforementioned mitosis, meiosis, cell division, and cell fusion, as well as other lesser occurrences that we might add, are not among them. For their part they seem to have been designed to do exactly what they do and nothing else. They are about reproduction and to the best of our knowledge have nothing to do with the survival of the parent cells. And unless some nonreproductive purpose can be demonstrated, natural selection cannot be saved this way.

NATURAL SELECTION ACTING ON GROUPS OF CELLS

The second option is that natural selection acts on groups of cells. In this case, destruction of the parent cell is irrelevant as long as the collective is advantaged by the generation of new cells. Reproduction is of value to the *community* of cells, not individual cells. And this is certainly the case. Reproduction is, as we have said, critical for the group's survival, for its continuation. The individual cell is sacrificed on the altar of the group, and this can be said to be the consequence of natural selection. In this case, the adaptation produced by natural selection would be reproduction itself, an adaptive property not of individual cells but of communities of cells.

But how can this be? Communities of cells would have to be invested with properties beyond the composite properties of their individual members. There would have to be communal properties on which natural selection acts. We will consider this possibility later, but for the moment, even if such properties exist, they are irrelevant to cellular reproduction. This is because it takes place and only takes place within individual cells or between pairs of cells. The properties of reproduction are those of individual cells. Though communities of cells are affected, cellular reproduction is not a community property. Simply, communities of cells do not reproduce.

NATURAL SELECTION ACTING ON PARTS OF CELLS

This brings us to the final possibility—natural selection acts on *parts* of cells, on their contained structures and molecules. This was discussed in chapter 4 in regard to a gene-centric view of evolution. We can extend this idea to any and all parts of the cell. For example, we can say that the mitotic spindle or the mechanisms of cell fusion and division are *themselves* subject to natural selection. Since, unlike the cell, they are not destroyed during reproduction, if an adaptive advantage applied to them, we could say that their evolution (as parts) was the result of natural selection.

However, as with a gene- or DNA-centered view of evolution, this is not possible. The natural selection of biological evolution takes place and only takes place between the environment and whole living objects. Parts of cells, however important, are neither living, nor can they somehow be considered wholes. In any event, how would such a selection work? How would natural selection choose among certain (inanimate) objects found *within* the living cell? Even if a component served as the basis for selection, the choice would be made not between different expressions of it in its own right but between the *living things* that contain the different manifestations. The cells, and not their contents, would be subject to natural selection, and as we know, they are destroyed in the process of reproduction. Whether we are talking about the whole cell (with its parts), or the individual cell (within the group to which it belongs), natural selection acts on the whole living cell. It acts on nothing else, not on parts of cells or groups of cells.

IN THE END

Each alternative, each attempt to account for the evolution of cellular reproduction in terms of natural selection fails:

1. The reproductive features of cells did not evolve for purposes of the survival of the progenitor cells.
2. Natural selection does not act on groups of cells.
3. Natural selection does not act on parts of cells.

Therefore, when all is said and done, however you look at it, destruction of the parent cell during cellular reproduction poses an insurmountable problem for natural selection as the agency at work here. By destroying it, reproduction cannot have provided an adaptive advantage to it. If we nonetheless insist that it must, we leave ourselves with nothing more than an oxymoron—the *destruction* of the progenitor cell provides it with an advantage for its *survival*.

And so, in destroying an existing cell to produce a new one, cellular reproduction presents an overwhelming difficulty for natural selection and hence for Darwin's theory. The conclusion just cannot be avoided that it is not the agency responsible for the evolution of the mechanisms of reproduction at the cellular level. When we look beyond the cell and examine the whole intricate array of reproductive mechanisms in complex animals and plants, we find the same thing. I am not referring to the destruction of something but to the absence of adaptive purpose for the progenitors, for the parents. Natural selection cannot account for the evolution of the mechanisms of reproduction, everything and anything from the means of sexual selection, to conception, fertilization, embryonic development, and finally to the emergence of a new being. This is the subject of the next chapter.

Chapter 7

BEYOND REPLICATING MOLECULES AND REPRODUCING CELLS

Natural Selection and the Evolution of the Reproductive Mechanisms of Life

Sexual reproduction is the chef d'oeuvre, the master-piece of nature.

Erasmus Darwin, *Phytologia*, 1800, 115:103

Evolution produced two parallel systems of life in a single body. One, existing life, fighting for survival, and another, reproduction, which gives rise to new life. DNA replication and cellular reproduction aside, can we attribute the evolution of both to the same agency, to natural selection? Is it responsible for the evolution of the reproductive systems of plants and animals, for everything from the elaborate mechanisms of flowering plants to the equally intricate reproductive systems of mammals, in the same way that it is thought to account for life's somatic embodiment?

As with cellular reproduction, the most direct way to pose this question is to ask whether life's reproductive features are of survival value to the creatures that express them, *the parents*. If they are, then there is a basis for claiming that they are the product of natural selection. If they are not, we are left to ask on what basis they evolved. To make this assessment we will use the reproductive process most familiar to us: mammalian reproduction.

THE FIRST PHASE OF SEXUAL REPRODUCTION

As I have explained, essentially all animals and plants, including, of course, mammals, are the product of sexual reproduction. We can think of sexual reproduction as occurring in two phases, each with a distinct end point. The first ends with the mixing of the parents' genetic material and the second with the realization of a new organism made latent by their union.

In mammals, the events of the first phase take place in the following sequence:

- *Meiosis.* The testes of the male and the ovary of the female manufacture special "germ" or reproductive cells by meiosis, a form of cell division that produces cells that contain only a single copy of the parent's genes (haploid cells), rather than the usual two (diploid cells).
- *Maturation.* The haploid cells produced by meiosis develop into mature eggs and motile sperm.
- *Sexual attraction.* In the meantime, external to the animal, a sexual attraction occurs between potential partners. This enticement is driven by appearance and behavior, as well as hormones, and culminates in the selection of a mate. Darwin referred to this as "sexual selection" to distinguish it from natural selection (see section III, "Sexual Selection").
- *Sex.* Sex or coitus takes place. The male places his penis (erection) into the female's vaginal canal, and the penis releases its sperm as the consequence of a back-and-forth motion (ejaculation).
- *The journey of the sperm.* Due to its motility, the released sperm travels from the vaginal canal to the uterus, where it finds an egg previously released from the ovary.
- *Fertilization.* In the uterus, a sperm fuses with the waiting egg.

So ends the first phase. Outside of sexual attraction, four major organs are involved: 1. the female ovary that makes the egg, 2. the male testes that makes sperm, 3. the penis that ejects the sperm into the female's vaginal canal, and 4. the vaginal and uterine canals of the female that carry the sperm to the waiting egg. There are also a variety of adjunctive structures, such as seminal canals (vas deferens) that carry sperm from the testes to the penis and oviducts (fallopian tubes) that carry eggs from the ovary to the uterus. All of this is elicited and coordinated by the nervous system and various hormones and is the consequence of biological evolution.

THE SECOND PHASE OF SEXUAL REPRODUCTION

In the second phase, the organism is created from the fertilized egg. This culminates in the birth of an autonomous but still immature being. The events that lead from the fertilized egg to birth occur exclusively in the female in the following order:

- *Duplication of deoxyribonucleic acid (DNA).* The fertilized egg copies its DNA.
- *Cell division.* The fertilized egg divides in half (asexual reproduction), and its DNA molecules are distributed equally to each daughter cell.
- *Blastocyst.* DNA duplication and cell division continue until a fluid-filled ball of some seventy to a hundred cells called the blastocyst is formed.
- *Implantation.* The blastocyst is implanted in the lining (endometrium) of the female's uterus.
- *Placental circulation.* Over time a special (placental) circulation between mother and fetus develops. Together implantation and the placental circulation provide a safe and nurturing haven for the developing embryo.
- *Development.* DNA duplication and cell division continue, being repeated over and over again until the organism is fully realized. This proliferation of cells is directed. It forms and shapes the body, and the cells that are generated differentiate into diverse types that give rise to the body's various tissues and organs.
- *Birth.* At some point the new organism is sufficiently developed to survive independently. By this I mean that it is able to circulate blood,

> breathe, regulate its own body temperature, ingest and digest food, and in most cases (but not humans) ambulate. In any event, at a specified time, driven by hormones, the new creature is pushed out of the mother's body by uterine contractions and vaginal dilation, and the baby is born.

This is just the barest sketch of these events. In addition, there is enormous variety in the details of the reproductive system from species to species. While some events, notably DNA replication and cell division, are common to all multicellular organisms, others, such as the placental circulation, are specific to a species or group—mammals, in this case. Underlying each step in the process are enabling mechanisms. For example, for cells to divide, they must have mechanisms that tell them when and how to divide, as well as how to sort the DNA equally to the daughter cells. These are the mechanisms of the cell cycle, cell fission, and mitosis. Though we know a great deal about the events of sexual reproduction, one thing that is unknown, or at least not fully known, is enormous by any standard. It is how cell division actually leads to the form and structure of the new organism, what is known as development.

NATURAL SELECTION AND MAMMALIAN REPRODUCTION

In any event, each structure, each mechanism, each of these processes is the result of biological evolution. Can it all—or, for that matter, can any of it—be attributed to natural selection? As explained, to make this determination we must ask whether the events of reproduction from fertilization to childbirth, singly or together, are of value to the *parents* in their struggle for survival. If so, then the claim for natural selection's role in their evolution would have an observational basis. But they are not. Not one of the events outlined above nor any of their subsidiary mechanisms provides the parents with an advantage in their life struggle.

This is unmistakable. Indeed, it is hard to imagine how the features of mammalian reproduction have *anything* to do with the continued existence of the parents. They seem to have one purpose and one purpose only, the generation of a new being. We can construct a parallel list for any plant or animal of our choosing, with the same conclusion. Procreative activities are simply not of survival value to the parents. Remarkably, these products of a long evolution are not of adaptive advantage to the organisms that express them.

That this is so is not only unmistakable; it is also self-evident. Natural selection is about survival, while reproduction is about producing new organisms. How can something that is based on survival value, natural selection, give rise to features of life that are about something altogether different, something that has nothing to do with the survival of the individuals that express them, reproduction? And yet, as explained, since Darwin's time, it has been thought that natural selection *is* responsible for the evolution of the reproductive features of life.[1] For this to be true, reproduction cannot be exclusively about procreation, or natural selection exclusively about survival. As outlined in chapter 4, they would have to be intertwined. Reproduction would have to be about survival, and natural selection about reproduction, as well as vice versa.

A UNIVERSAL THEORY

From the most to the least general, Darwin's theory has three forms:

- Natural selection is responsible for the evolution of all the features of life. Its role is all-inclusive; it accounts for all of life's traits, in all species and without exception.
- More cautiously, natural selection accounts for most but not all of life's features. It may border on the ubiquitous, but there are exceptions.
- Most limited, natural selection accounts for some features of life, but the extent of its rule is left unspecified.

Alfred Russel Wallace was a proponent of the general, or "strong," form of the theory. He thought that the theory could only be true if it was universal. Negative examples would perforce prove it false. Darwin's doubts about the theory's universality grew over time, and it seems that he came to believe in the second version, the existence of exceptions. This was at least in part a reflection of his belief that the traits involved in sexual selection lacked survival value (see section III, "Sexual Selection"). The third proposition, the most guarded statement of the theory, though not usually associated with a particular individual, allows for natural selection but leaves unsettled how widespread and essential it is. Many observers seem to have flitted back and forth between the three views.

In any event, if either proposition two or three is correct, if there are exceptions to natural selection, then proposition one, the universal proposition, is of course false. This is an enormous problem because only the first proposition, the universal statement, is a proper scientific hypothesis. This may seem odd because the other two seem no less plausible, perhaps even more plausible because they allow for exceptions, and certainly there are exceptions to all kinds of generalizations. The problem is that a hypothesis that allows for exceptions cannot be disproven. For an idea to be scientific we must be able to show that it is false. Propositions two and three would remain unbowed even if natural selection were rejected for one feature or another, indeed, for the third, even if it was rejected for many.

This said, proposition one exacts a harsh discipline. As a matter of method, it cannot allow for exceptions, for reproduction or anything else. This means that if just one feature of life can be said *not* to have evolved as the consequence of natural selection (or cannot otherwise be connected to it), the theory must be false in general. In line with Wallace's point of view, it does not allow us to claim that the theory is true despite observations for which it is patently false. A valid exception declares its failure.

There is also the question of purpose. The purpose I am referring to is that of science in its search for universality in understanding nature. Science's clear and unambiguous mandate is to seek the broadest possible explanation for phenomena, not partial explanations, not theories that explain this but not that. As such, it attempts to develop theories that are able to account for all phenomena in the ambit of its concerns. Even though much of a scientist's life, certainly that of biologists, is spent considering small hypotheses about what he or she sees, the ultimate mission of science is broad generalization.

It is hard to imagine science's search for universality or the desire for generality not to include a phenomenon as broad and all-encompassing as the means and mode of life's evolution. If anything calls for a universal explanation, certainly it is the evolution of life. And of course this was what motivated Darwin and Wallace to develop their theory in the first place, however they and others eventually came to view it.

Evolutionary theory is like atomic theory. It proposes a basis for all life, just as atomic theory does for all matter. It does not merely seek to account for the evolution of this or that, for some things and not others, but for everything. Whatever Darwin came to believe, just as all material things are made of atoms, by its very nature his *theory* proposes that all of life's incarnations came into being as the consequence of natural selection.

PROOF

Despite science's call for universality, it is often thought that for a scientist to note exceptions to his or her theory is a virtue, a sign of being scientifically rigorous, of being careful not to overstate the soundness or unassailability of your viewpoint. But when exceptions are noted *without rejecting the theory*, this honesty may be faux. It may be a dodge, a coy evasion. "Yes, the exceptions must be dealt with, but all things considered my theory is correct nonetheless." This may have been what Darwin was up to when he expressed doubts about his theory's universality while at the same time retaining the belief that it was true.

And this is why some think that Darwin's theory is not a scientific theory at all. When challenged by negative evidence, it can devolve from proposition one into propositions two or three and escape falsification. To claim true scientific status, it must be stated as a universal proposition, falsifiable by a single valid exception. While this is a very stringent standard, given the enormity of the claim about life's evolution, a resolute, strictly observed, even an unbending criterion for acceptance seems perfectly appropriate.

Yet such a requirement may appear not only rigid but downright foolish, even irrational. Why reject a perfectly good theory just because of one or two failures of its predictions? If we can cite numerous affirming examples, cases where the theory seems to apply, it seems harsh and unyielding, as well as unrealistic to demand its rejection on the basis of a single or even a few failed instances.

Against this, the philosopher Karl Popper cautioned that however many facts support a theory and whatever their nature, scientific theories are necessarily and forever conditional. However well ensconced, they are *always* vulnerable to future tests or observations that show them to be false. Even if a particular theory has been up to the challenge on many occasions, and that as a consequence we have gained a great deal of confidence in its correctness, science can never accept it as being true beyond doubt, because the next test, the next observation may prove it false. According to Popper only *failed* scientific theories provide clarity about the character of nature. All we can know with any certainty about the world around us is what is verifiably false; what is untrue.

Still, if numerous observations support our theory, why should we meekly accept failure and allow a negative observation or two to trump all others? Why should we let our confidence evaporate on the basis of a few negative cases? Why should we have to reevaluate the meaning of affirmative observations, of which there may be many, when perhaps it is the negative observations, of which there may only be a few, that are in error?

Incongruously, Popper agreed. He said that it is often prudent to hold on to one's theory despite a failed test of its validity. He said that in the final analysis we should not be satisfied with a single piece of falsifying evidence as the basis for discarding a perfectly good theory. He counseled additional tests, additional examples of failure, additional contradictory evidence before we reject a theory that has much to recommend it.[2]

But how would this work? How many tests would be needed—two, ten, a hundred, how about a thousand before we accept that a particular theory or hypothesis is incorrect? What is the appropriate number? And how do we weigh falsifying observations that may not be equally dispositive? The fact of the matter, I would say the sad fact of the matter, in line with Kuhn's paradigms, is that conclusions drawn about the truth or falsity of a scientific theory depend on all sorts of standards and circumstances other than on unadorned reason. In fact, the basis for acceptance or rejection may not be scrupulously scientific or for that matter scientific at all. It might be based on nothing more than preconceptions and prejudices that represent little more than an emotional attachment to a particular point of view. As noted, such situations abound in regard to the theory of evolution.

Given this state of affairs, how can we truly and honestly determine whether Darwin's theory or any theory for that matter, especially one that is well respected, is false? Unfortunately, there is no easy answer to this question. Each situation must be taken on its own. This brings us to the current circumstance. How are we to view Darwin's theory if natural selection does not fit the bill for life's reproductive features? What are we to make of it?

We have three choices. All are methodological. We can simply reject the theory. We can say that since there is clear proof that it does not apply to reproduction, that it must be false in general. Or we can soldier on looking further, hoping that future observations will clarify matters and save the theory along with its universality. Finally, we can reject it as being universal, but not otherwise. Reproduction may be the exception that proves the rule.

As already suggested, I am wedded to the second path—soldiering on in an attempt to save the theory, to show that after all is said and done natural selection accounts for the evolution of life's reproductive features. This approach not only has the advantage that it may be true, but also, if the effort fails, we will gain confidence in its inescapable rejection. Furthermore, by following this path, we always have the option of falling back on the third choice—reproduction as the exception to a generally applicable rule.

My course of action in what follows is based on two mindsets; one is objective and the other subjective. The objective frame of mind is the insistent search for universality in nature, while the subjective attitude is a persistent, not to say obstinate resistance to the theory's rejection whatever the evidence. Either approach can be easily damned. The theory may be said to be too important to be accepted with an undemanding acquiescence to a counterfeit universality or, alternatively, too important to toss out on the basis of flimsy negative evidence. If at times in what follows I seem to be gilding the lily, piling on when the point I am trying to make is already clear, please bear with me. My purpose is to transgress or at least forestall damnation by being as unmistakable and unambiguous as I can about the conclusion being drawn.

REJECTION

Any attempt to resolve the dilemma that life's reproductive features pose for classical Darwinian theory must be made in light of its central claim about the consequences of natural selection. As explained, for natural selection to be responsible for the evolution of a feature of life, whatever that feature may be, the following must apply:

- The evolved feature must serve, or in some way be related to, an adaptive function or property.
- Adaptive functions or properties, whatever their particular nature, must at least to some minimal degree improve the likelihood of an individual surviving environmental challenge.

In this, survival is not merely the sine qua non of natural selection; it is its sole selective criterion. There is no other, and the theory proposes no other. With this in mind, and as explained, we can conclude that natural selection is the agency of reproductive evolution only if the evolved mechanisms and processes offer an adaptive or survival advantage to the individuals that express them, once again, the prospective parents.

Also once again, this does not seem to be the case. The events of reproduction seem to have nothing, at least nothing positive, to do with the survival of the parents. If natural selection is about survival and reproduction is about creation, how can natural selection be the basis for the evolution of life's reproductive features? Survival does not create anything.

Certainly this was known to Darwin, Wallace, and their contemporaries, as well as to the many biologists who have thought seriously about evolution in the 150-some-odd years since publication of *On the Origin of Species by Means of Natural Selection*. And yet, it has been widely if not universally believed that natural selection *is* responsible for the evolution of the reproductive features of life. We considered the affirmative reasons for this belief in chapter 5, but there are also some powerful negative reasons:

- Rejection of natural selection as the agency of life's reproductive features would leave us without an explanation for their evolution. Science currently has no other acceptable account.
- If we exclude reproduction from its purview, the theory of natural selection would no longer be universal, and, as explained, universality is imperative if the theory is to have scientific merit.
- Given the vast diversity and intricacy of reproductive mechanisms, if we reject natural selection as the agency of their evolution, the theory would be decimated by multitudes of defections and apostasies and would be unable to withstand this abandonment.

THREE AFFIRMING POSSIBILITIES

Given powerful reasons, both affirmative and negative, for accommodating reproductive evolution to Darwinian theory, attempts to keep it in the fold should come as no surprise. Often the inference is drawn from one or another supportive observation, from a particular instance, while at other times it is merely notional. Crucially, belief in the theory's hegemony and the inclusion of reproduction is *not* the result of deductive reasoning designed to validate the conclusion. By and large these efforts at accommodation have followed one or another of the three lines of attack that have already been noted for cellular reproduction:

- However it may seem, reproductive features evolved in the parents for purposes of their survival.
- Natural selection acts on groups of organisms, as well as on individuals. And finally,
- Natural selection acts on parts of cells and organisms.

GROUPS AND PARTS

Let's consider the second and third possibilities first. They are analogous in that both claim that natural selection acts on them *in their own right* instead of or in addition to the individuals that comprise or contain them.

This is simply not possible either for life's reproductive or somatic features. As for groups, central to Darwinian theory is the idea that groups of organisms, not individuals, evolve. We live and die as we are, unaltered individual beings. Only groups evolve. As explained, this is the key difference between Lamarck's concept of use and disuse and Darwin's evolution as descent. And yet by its very nature, natural selection can only act on individual organisms. Groups are assemblages of members and as such have no independent material incarnation. Viewed as entities unto themselves groups are abstractions, and natural selection can only act on material bodies.

As for the parts of living things, natural selection can only act on them as ordinary physical objects. It can only apply to their *biological* properties when and where they are manifest, and that is in the whole cell or organism. Their evolution is unconditionally and categorically tethered to the survival of the whole object. That is where natural selection acts. For example, how could natural selection affect the genetic code, that is, the information content of DNA in its own right, absent the context of the whole organism?

THE ABSENCE OF ADAPTIVE ADVANTAGE

There are two variants of the claim that reproductive features evolved to help ensure the parents' survival. First, as already noted, reproductive features may have evolved for nonreproductive purposes, for purposes related to survival. In this case, their reproductive functions would be ancillary by-products. We will talk more about this in the chapters on sexual selection, but for now let us consider the other form of the claim. Simply and directly, reproduction in and of itself has survival value for the parents.

To validate this view, it is not sufficient to provide examples where reproduction is in fact of survival value to the parents. For the claim to be true, we must show that it is true for each and every reproductive feature in each and every species. In this case we cannot generalize from examples. We cannot inductively infer general rules from specific examples without first assuming that our examples are exemplary, that they apply in general, when this is after all what we are trying to find out.

We would also have to show that such advantages as there may be are of sufficient weight to conclude that the mechanisms of reproduction in all their variety and complexity evolved for this reason. Given the centrality of reproduction to biological evolution, the effect on survival would not only have to be significant; it would have to be essential.

Finally and most critically, it is the circumstances in which such adaptations are *not* seen that are controlling, as negative evidence always is. And the absence of adaptive value turns out to be commonplace. Though adaptations associated with reproduction useful to the parents are seen, many, and perhaps most, reproductive features provide *no* such advantage.

This point can be made with one simple illustration. In a great many species, vertebrates, invertebrates, and plants alike, one or both parents are *not* present when fertilization occurs (or during the subsequent reproductive unfolding). How can *the mechanisms of reproduction* provide an adaptive advantage to absent parents? If we had to fashion a general rule about such an advantage, based on the preponderance of the evidence, we would have to say that the mechanisms of reproduction provide none.

Not only that, but when there is an effect, it is often, if not usually, negative. Having children can be disadvantageous, even disastrous for the parent. Extreme cases are found in certain insects, where the male does not survive sex. For example, in the honeybee, the male's (drone) sexual organs are left inside the female, and it dies as a result of the wound. Or the female praying mantis devours the male after sex. Closer to home, the female mammal is often placed in great jeopardy bearing and giving birth to children.

The plain fact is that most animal and plant species live and die, live well or die harshly, whether or not they procreate. In most cases at whatever level and whatever aspect we examine, from courting to birth to everything in between, the parents gain no adaptive advantage from their participation in the reproductive process.

CHILDREN SERVING THE NEEDS OF THEIR PARENTS

But more often than not, it is not in what happens before birth but in what happens after that an adaptive advantage is claimed. The contention is that parents profit not in the act of *having* children but in what happens after they are born. Having been born, they serve their parents' needs.

In the first place, a growing population improves the chances of a particular individual surviving a predatory threat simply as a matter of numbers, of statistics. The larger the population is, the less likely that you will be the target. A selection based on numbers is also biased in favor of the parents. The young are usually easier targets, slower, less powerful, less skillful or crafty. Beyond such advantages, offspring also often serve their parents in affirmative ways. Among animals, they may obtain food or act as defenders (soldiers). Many insects are fated to be soldiers (to fight) or workers (to obtain food). For humans, even in developed countries, children sometimes work to help support the family, and in the underdeveloped world, their work may be essential for the survival of its members. Of course in all human cultures, it is primarily young male adults who fight wars.

Even so, whatever the species and in whatever ways, the presence of offspring can only be beneficial *up to a point*. That point is when they *compete* with their parents for limited resources or space. This is critical for plant life. New plants compete with their parents as well as each other for soil and sun. Needless to say in such circumstances, having progeny affords no selective advantage to the parent. We can say that the parent has made a poor bet, a counter-adaptive gamble.

In the end, we cannot deny the fact that *in most species* the parents gain no adaptive advantage either from reproduction itself or from their progeny after they are born. If, as said, a judgment must be made, it must be negative, governed by the many negative cases. In the end, all three possibilities, survival value to the parents, group selection, or the selection of parts, fail. Whatever our wishes, we are drawn ineluctably to the conclusion that natural selection is *not* responsible for the evolution of the reproductive features of life.[3] But the problems for natural selection do not end here. We now turn to another difficulty exemplified by Darwin's theory of sexual selection.

SECTION III

SEXUAL SELECTION

Chapter 8

IRIDESCENT BRILLIANCE

Darwin Worries about Peacocks and Finds Sexual Selection

The Pride of the peacock is the glory of God. The lust of the goat is the bounty of God. The wrath of the lion is the wisdom of God. The nakedness of woman is the work of God.

William Blake, "Proverbs of Hell,"
in *The Marriage of Heaven and Hell*, 1793

Darwin's world was filled with many marvels of nature, and one of his most extraordinary talents was seeing beyond their mere appearance. Unlike most naturalists of his day, he did not just describe remarkable things; he constructed his general theory of evolution from them.

None of the objects that caught his attention was more disquieting to him than the splendid plumage of the peacock with its many shimmering eyes. What conceivable purpose could these strange yet glorious feathers serve? If anything, they seemed counterproductive. If they were for flying, they seemed clumsy and unwieldy. If they were for protective coloration, they seemed to draw attention, not provide camouflage. And what of the enormous fanlike display? Though beautiful, it seemed an awkward hindrance at best.

These were not questions of idle curiosity to Darwin. His theory of evolution seemed to stand or fall on the answers. Whatever his personal beliefs, a central feature of his theory was the requirement that *all* evolved features serve or be related to others that served an adaptive purpose for the organism and that as such aided the *utilitarian* objective of survival. If the peacock's plumage was merely *decorative*, if it provided no adaptive advantage, if it was of no help in surviving life's challenges, then what had spurred its evolution?

There was one rather obvious clue. Only the male peafowl sported this flamboyant and luminous plumage. The peahen, the female of the species displayed none of it. Its feathers were plain, its appearance lackluster. And this, Darwin appreciated, was common among birds. The male tended to be colorful and ornamented, while the female was, relatively speaking, drab and dowdy.

As a consequence, it seemed that the peacock's plumage had something to do with sex. And bird feathers were not alone in this. There were sex-related traits of all sorts in as many species, which, like peacock's feathers, were not part of reproduction per se. What to make of such features? Unsurprisingly, Darwin thought they might serve an adaptive purpose that was related to their sexual dimorphism, that is, their presence in one sex, but not the other. But what was that purpose?

SEXUAL SELECTION

It was common knowledge that some of these features had to do with choosing a mate. With the iridescent brilliance of the peacock's plumage at the head of the class, Darwin appreciated that there was a *sexual selection* in which certain ancillary, known today as secondary, sexual characteristics functioned as sexual attractants, as means of attracting mates.

And such means were critical. If evolution depended on reproduction, and if it could only occur pursuant to the selection of a mate—and this was obviously the case—then the ability to obtain a mate was as critical as natural selection, as important to evolution as survival itself. Participation in evolution required two things—surviving to sexual maturity *and* finding a mate. It was with this understanding that some twelve years after the publication of *On the Origin of Species* in which the idea was briefly noted that Darwin fleshed out a theory of sexual selection in *The Descent of Man and Selection in Relation to Sex* (1871).

He described two types of sexual selection that were analogous to natural selection's interspecific and intraspecific selection, except that in this case selection was not between or within species but rather was between or within genders. Between genders, there was an enticement, a chemical, a physical, and—for species like humans—a psychological and culture-laden sexual attraction, as well as an accompanying response in the attracted sex in deciding whether or not to accede to the blandishments.

The attraction as well as the choice of partner could be the provenance of either sex or, for that matter, both. As said, in birds, the beauty of the male is thought to attract the female, while the attracted female selects her mate. In modern human society (as opposed to traditional human societies), the female often both attracts and chooses the male. In many species the attraction is mutual, as is the choice of a partner.

Within genders sexual selection involves competition between members of the same sex for the favor of members of the other sex. Darwin's favorite example was nonlethal combat between males as a preface to sexual intercourse with a female. By winning the contest, the successful male established a preferred place and in some cases dominion over sexual activity. In the discussion of sexual selection that follows in this and the following chapters, the emphasis will be on inter-gender selection, on the attraction between the sexes.

THE NATURE OF SEXUAL SELECTION

Beyond the commonality of choices, within or between species or within or between genders, selecting a sexual partner and natural selection are as different as night and day. Like the relationship between reproduction and natural selection, sexual and natural selection seem antithetical. Consider the following:

- Unlike the breadth of causative environmental agencies for natural selection, there is only a single cause for sexual selection—the prospective mate.
- The causes of natural selection elicit all manner of response in the affected organisms, whereas in sexual selection there is only one, the desire for sex.

- Unlike natural selection, sexual selection occurs wholly among living things; there is no equivalent to the inanimate physical forces, such as ambient temperature, that are important causal agents in natural selection.
- The environmental cause in natural selection is unwanted, a threat; in sexual selection, a mate is sought.
- A negative outcome of natural selection may be death; whereas a negative outcome of sexual selection only means that the excluded organism does not procreate.
- The environmental threats (e.g., temperature and predation) that are the cause of natural selection are general and apply to all species, whereas the attractants of sexual selection are sex and species-specific. For example, in the human female, a pretty face; a well-turned ankle; prominent breasts, hips, and buttocks may be features of desire to the male, and bulging biceps, broad shoulders, and strong chins may be sexual attractants to the female, but none of this has any meaning to birds, worms, or flowering plants. And finally,
- While somatic traits are the consequence of natural selection, none are devoted to it. That is, there are no natural-selection genes or organs. However, there are a multitude of inherited and learned traits dedicated to sexual selection, such as beautiful feathers and sexy movements.

THE UBIQUITY OF SEXUAL SELECTION

Given these facts, is sexual selection a general principle in the same way that natural selection is? For instance, does it occur in all sexually reproducing organisms? In one sense, it does; indeed, it must. At the inception of sexual reproduction, in the coupling of male and female cells, in the fusion of sperm and egg, a sexual selection is made, a partner is chosen. With some rare exceptions (such as haploid male [drones] honeybees), this selection is a necessary prerequisite for sexual reproduction. Reproduction presupposes its occurrence.

This, however, is not the sexual selection that Darwin had in mind. In the fusion of egg and sperm, the choice of mates is a matter of chance between what is usually a large number of possible mergers. The choice is random, while Darwin's sexual selection is anything but. It is an affirmation of a *particular* individual's desirability. If fusion between male and female cells, between sperm and egg is ubiquitous and universal, Darwin's sexual selection, though common enough, is far from it.

The enormous and diverse group of land plants, the flowering plants or angiosperms, present an ineffaceable exception. Though the color and form of flowers are important sexual attractants for cross-pollination (when one flower's pollen [sperm] fertilizes another's egg), it is *not* the opposite sex that is attracted. As every grade-school student knows, it is insects—bees, butterflies, and the like, as well as nectar-feeding birds, such as hummingbirds.

Insects and nectar-feeding birds act as the plant's sexual servant, carrying its pollen from flower to flower. And so, the attraction in this case is not between male and female plants but between a plant and another species, an animal species at that. What is more, the attraction is not sexual, at least not for the insect or bird. All it is doing is feeding on the flower's nectar. It is attracted not to sex but to food. As a consequence, though the attraction has a sexual purpose for the flower and of course floral reproduction provides more flowering plants and consequently more nectar for insects and nectar-feeding birds, and therefore the relationship between the two is symbiotic, it occurs without sexual purpose for the animals. It is a kind of unintended farming, unintended because the pollen is carried adventitiously.

WALLACE'S OBJECTIONS

Wallace had difficulty with Darwin's theory of sexual selection.[1] It was not that the idea that mates were selected disturbed him. That was obvious enough. We are sexually attracted to certain people, but not others, and this attraction is necessary for sexual desire. What troubled him was that the traits involved in this choice did not seem to be the product of natural selection.

Though natural selection and sexual selection are corresponding events of commensurate importance, and together account for the occurrence and course of evolution, they otherwise appeared to be fundamentally and self-evidently different. Natural selection was about survival, while sexual selection was about choosing a sexual partner, about procreation. And procreation is not survival. As such, the features of sexual attraction could not, it seemed, be the products of natural selection. To Wallace, this was a great threat to his theory. If it was true, then the theory of evolution by means of natural selection was not a general or all-inclusive theory and had to be rejected. As far as he was concerned, natural selection had to apply to the evolution of *all* of life's characteristics, or it was mistaken for them all. Today this view is called *Adaptationism*.

Darwin seemed to be arguing against his own theory. For the theory to be correct, and Wallace thought it was, natural selection *had* to be the agency responsible for the evolution of features of sexual attraction. For this to be so, they had to have evolved for purposes of survival. There was no choice. Wallace's solution was that the sexual function of these features, though undeniable, was incidental or secondary. With this in mind he set out to show that features associated with sexual selection evolved for ordinary somatic adaptive purposes and only subsequently or incidentally acted as sexual attractants. He was quite successful, but in the end he could not make a general case. He could not attribute each and every attractant feature to some parallel somatic purpose, at least not to his own satisfaction. Some, like peacock feathers, seemed to serve none.

And so, Wallace ended up rejecting his own explanation though he never seems to have acceded to Darwin's view. Anyway, the theory of sexual selection was left standing, a scab on the theory of natural selection. When it was discussed, not least by Darwin himself, it was usually in terms of a fascinating coterie of features.[2] The problem they posed for natural selection was usually ignored. Still, ignored or not, these important features appeared to have evolved by some other means.

POWERFUL, BUT OBSCURE

The mechanisms behind natural selection as it is found in biology are both the living and inanimate forces in the environment that act on organisms. By studying their relationship to the somatic features of life, we can come to understand the mechanisms that underlie the feature. For example, we can understand how temperature affects biological structure, reactions, and behavior, or how mechanically and chemically a predatory attack is carried out, with teeth, muscle, poison, barbs, sticky substances, and so forth. We are able to explain such things in straightforward mechanistic terms by connecting the environmental causes of natural selection to its effects.

But when we attempt the same kind of analysis for sexual selection, when we look to the environment to grasp the mechanisms that underlie the evolution of these features, as well as those that underlie reproduction more broadly, we find nothing. There are no environmental forces, no physical or chemical forces, no biological forces that drive the selection of more sexually attractive embodiments over less attractive ones, or that distinguish those seeking sex from those who are disinterested. The basis of sexual desire is obscure.

Certainly the attraction exists, and we can explain many of these events in physical and chemical terms, and for species where it is relevant, in psychological terms, but as best we can tell, the only reason for the existence of a sexual attractant and the response to it is sexual selection itself. There is no underlying physical basis or other reason for it, such as ambient temperature. This burdens us with another tautology. A sexual attractant is a sexual attractant, simply because it is a sexual attractant. This is not to say that it is without purpose. Of course it leads to sex. But there seems to be no underlying mechanistic basis for the attraction beyond the attraction itself.[3] The confusion deepens as we now turn our attention to two of the most puzzling properties of sexual attraction: the importance of small differences and beauty.

Chapter 9

ON THE IMPORTANCE OF SMALL DIFFERENCES

The Basis of Sexual Attraction

> *I can never bring you to realize the importance of sleeves, the suggestiveness of thumb-nails, or the great issues that may hang from a boot-lace.*
>
> Sir Arthur Conan Doyle, "A Case of Identity," in *The Adventures of Sherlock Holmes*, 1892

Though the body's internal workings play an important part in generating sexual activity; for example, sex hormones may signal receptivity in a potential partner, sexual attractants are by their very nature external manifestations. They are found in how organisms look, in evocative smells, in touching, as well as in certain movements and behaviors. The reason is simply that a sexual attractant must be perceived, seen, or otherwise sensed by a potential sexual partner.

With this in mind, and given that stimuli produce responses, it may seem nonsensical to say that while the presence of attractants and responding partners are necessary for sexual selection, they are not sufficient to produce it. I do not mean that sex may not occur because circumstances get in the way, for example, an attack by a predator, foul weather, or a bad meal in a fancy restaurant, but that even under the most propitious circumstances the mere presence of an attractant and the desire to respond sexually does not permit an affirmative, nonrandom preference among potential partners. A choice just cannot be made.

The reason is that with some minor exceptions, individuals of the same sex in the same species express *the same* attractant features. Selection would have to occur between different incarnations of the same thing; that is, between identical objects or processes, and as such could only be random. There is no basis for choosing one partner over another. A pheromone (a volatile chemical sexual attractant) released into the air in roughly equal amounts by members of one sex would attract members of the opposite sex, but indiscriminately. They would be attracted with the same intensity to all those who release the substance.

As with the fusion of sperm and egg, this is not Darwin's sexual selection. It is about the *relative* attractiveness of different individuals, not a random choice among them. If all attractants within a species were identical from individual to individual, if they were doppelgangers, selection could only occur as a matter of chance. For one individual to be favored over another, their attractiveness must differ.

For this to be the case, they must have different embodiments, different sexual valences or values that are noticeable to members of the opposite sex. For pheromones, this would mean that the relevant molecule would have to be released in meaningfully different amounts from one individual to another. An attractant cannot just be an attractant. A smell cannot just be a smell; a feather, a feather; plumage just plumage; a nose cannot just be a nose; nor a leg, a leg; and actions however similar—strutting, dancing, posing, or smiling—cannot be identical from individual to individual. There must be something more or something less in form, size, color, shape, motion, amount, rate, level, or, for psychological attractants, affect in different individuals sufficiently distinctive for the difference to be perceived by the other sex. Only then can sexual selection take place as an affirmative choice.

BEING DIFFERENT IS NOT ENOUGH

But being different is still not enough. The differences must be perceived as being consequential. That is, they must make one organism more or less desirable, more or less attractive, or more or less appealing than another. In one way or another the attractant feature must be more or less crude, more or less subtle, more or less elaborate, more or less insinuating, not to mention stronger, larger, weaker, smaller, more or less overpowering than those of other individuals. Whatever the comparative basis, attractants must differ in some significant *affirming* respect in the chosen individuals compared to those that are not chosen.

In an example that dates back to Darwin, the male bowerbird builds an elaborate bower to attract the female. Some males do a better job than others; that is, they produce a more enticing bower, and are chosen by the "shopping" female. In a similar vein, Darwin talked of the size and embellishment of the stag's antlers. He thought that larger, longer, and more elaborated antlers were more desirable to the female. There are also dances of attraction, from the seemingly odd movements of insects and arthropods, to the sexually infused gavotte and tango of humans that are more alluring when performed by some individuals than others.

And finally there are utterances, from guttural noises to beautiful birdsongs, and for humans, conversation. Of course sounds are not sexual attractants per se, but when they are, whether harsh or mellifluous, their sexual valence depends on the emotion being conveyed and how it is received, and this varies from individual to individual. With humans, the same words can be perceived as alluring, wishy-washy, foul, or even abhorrent depending on how they are said, who says them, to whom and under what circumstances.

THE IMPORTANCE OF SMALL DIFFERENCES

Though such differences can be powerful drivers of sexual activity, commonly they are anatomically and functionally trivial. Small, relatively insubstantial differences in sexually charged traits, that have little if any bearing on the trait's functionality, often distinguish sexually deficient incarnations from commanding attractants. This is so, whether we are talking of bowers, antlers, dance, song, or words. Certain differences, however insubstantial otherwise, can be exceptionally potent as sexual attractants and hence as promoters of sexual activity.

Human faces offer a good and thoroughly mystifying example. Of course, all faces have the same anatomical features: eyes, noses, mouths, teeth, cheekbones, chins, foreheads, and so forth. And the appearance of human faces is roughly the same from individual to individual, and is obviously different from that of other primates, no less dogs or cats. Yet, it is not the mere presence of human facial features that is sexually attracting but *fine differences* between individuals. Facial beauty depends on particularities of size, form, and shape, as well as on spatial relations and geometric balance between the parts and how they are put together to form the overall topology of the face.

By nonaesthetic or utilitarian standards, these differences are almost always anatomically and functionally inconsequential. They are of no practical significance. But however inconsequential, however insignificant they may be, they can make one individual desirable, attractive, pretty, or handsome, and another unappealing, unattractive, plain, or homely, even repugnant. This is the power of small differences. It applies to features of sexual attraction in general. Whatever the basis for the attraction, whether bowers, dances, antlers, or faces, and of course peacock feathers, small differences can be of enormous selective consequence, enhancing the chances of a particular individual being chosen or alternatively lessening, even eliminating it.

SMALL DIFFERENCES AND NATURAL SELECTION

Whether appealing or repulsive, these subtle distinctions are not merely descriptive of a given individual; they are inherited. They are the result of genetic and presumably evolutionary causes. In humans the similarity of minor attributes in facial appearance is perhaps the most commonly understood basis for concluding that there is a close familial (genetic) relationship between individuals. Dating back centuries before Gregor Mendel and genetics, all sorts of court documents attest to the fact that similarity in appearance is strong, presumptive evidence of propinquity, proof that two individuals are closely related, that they share a common genetic heritage.

Likewise, small *differences* in appearance can indicate the opposite, the *absence* of close family ties, even different tribal or racial origins. Anyway, whatever the relationship between individuals may be, some of these small, inherited differences—facial and otherwise—serve as the basis for evaluating the sexual desirability of a potential mate. This has two seemingly contradictory implications, insignificant by one standard and vital by another.

What is insignificant is that genes (most likely some sort of combination of genes and regulatory complexes, not single genes) do not merely specify nose, chin, zygomatic arch (cheekbones), and so forth, they detail their shape and size, including all sorts of slight functionally meaningless gradations. What is vital is that these trifling genetic differences can have considerable sexual content and power. A particular anatomical form may be perceived as beautiful or ugly, when by objective standards it may differ little from another that is either seen as the opposite or as nondescript. In the same way, a bodily movement may be seen as sexy in one individual and clumsy in another, or a sound may be beautiful in one and grating in another, and on and on, small difference upon small difference, inconsequential by one standard and vital by another.

However small and functionally insignificant these inherited differences may be, they are presumed to be the consequence of natural selection. But how would this work? How would natural selection act to produce the particular feature in the absence of a significant utilitarian disparity? If the only distinction between them is related to sexual selection, if the desirable feature provides no adaptive advantage related to survival, then how could natural selection be the agent of their evolution?

DIVERSITY AND SAMENESS

There is another problem. Natural selection has two conflicting roles in evolution. On the one hand, it helps establish life's *variety* by selecting among individuals and species best able to survive in *different* environments. Most broadly, some survive on land, others in water, some do best underground, and yet others up in the air, while some survive at high temperatures, others at low temperatures, and finally some are better able to defend against predators or to obtain food in a given environmental situation than others. In each case, natural selection serves to establish useful (adaptive) ends that are distinctive, diverse, and divergent from species to species.

But it also works in the opposite way—it acts to produce *fidelity*. In this case, natural selection selects for the faithfulness of a feature to some particular end uniquely suited to a common environment. In this circumstance, it seeks not different ends but a shared ideal. Through its actions, evolution progresses toward some *optimum* incarnation of a particular trait. Sometimes acutely, but customarily over long periods of time, generation by generation, natural selection works its will, progressing toward, though never reaching, a most adaptive form. In any event, the tendency in this case is to produce sameness, not variety.

For the evolution of life's somatic features, these two roles exist side by side, in perpetual conflict, one that seeks diversity due to differing environmental demands, and the other that seeks an optimal incarnation due to the constancy of those demands. The character of what evolves is the consequence of this conflict. Things are different for sexual attractants. In this case; selection is based *only* on environmental sameness, on fidelity. The reason is that in any given species, sexual attraction takes place in a common or comparable environment for all instances of its occurrence. This may seem incorrect and obviously so. Surely, the environment in which sexual selection takes place varies from occasion to occasion. The couple seeking sex is surrounded by the unique circumstances of the moment.

True enough, but this is not the environment I am referring to. Although the circumstances of the occasion and the moment are certainly important, they are merely adjunctive or tangential. The core natural environment of the encounter is the potential sexual partner in his or her own right. The environment outside the relevant pair certainly affects what occurs, but the sole necessary and only sufficient environmental ingredient is that embodied in the prospective sexual partner. Needless to say without his or her presence, there is no sexual encounter, no selection, no sex.

Though the partner may be different in each encounter, given the sameness of the circumstance—the presence of a prospective partner, selection, if operative, would act to generate sameness, to produce the most effective instantiation of an attracting trait. This would mean that all individuals of a species would tend to become more attractive and equally so. As a consequence and counterproductively, the distinctions needed for selection would fade over time and eventually disappear, along with the basis for selection.

AGAINST AN ADAPTIVE IDEAL

What is needed for evolution is not sameness but variety. Different forms of a trait cannot be equally adaptive. One must always possess some selective advantage over the other. For life's somatic features, this difference has intrinsic and extrinsic causes, in particular mutations and a changing environment.

On the other hand, for sexual selection, where the relevant environment is the sexual partner, the setting is unchanging. In this case, mutations aside, if natural selection is the agency of selection and different individuals are attracted in roughly the same way by the same attractants, evolution would work toward a single most desirable, most appealing embodiment throughout the population, and as a result as the generations passed, sexual selection would disappear. It would be lost in a homogeneous display of sexual attractants with equal valance and value among individuals, with no meaningful choice being possible among them. They would be true doppelgangers.

Of course this is not what happened. Differences in the attractiveness of individuals was not eliminated but rather was maintained. The attractiveness gap was not closed but was preserved. Features, whether more or less appealing, were inherited and continued to be responded to as such. Distinctions were not smothered by evolution but were maintained by it.

This leads to another contradiction that carries us even further from an adaptive ideal. Procreation is not limited to organisms that express the ultimate, the most sexually desirable features. We know from our own experience and from observing other species that reproduction is not limited to the most attractive creatures. If it were, they would come to dominate the population, and consequently variety in attractiveness would disappear. We would all be gloriously attractive. Whatever else we can say about the evolution of sexual attractants, the agency that gave rise to them is able to produce and maintain consequential distinctions between individual incarnations of the relevant features *despite* their being expressed in ostensibly the same environment (once again, that being the presence of the potential partner).

ON THE SMALLNESS OF DIFFERENCES

The most challenging problem for natural selection in all this is that however important and powerful small differences may be in our sexual lives and presumably those of other species, what makes one incarnation more desirable than another is obscure, even incomprehensible. It is not that we cannot describe the differences or what attracts or repels us. That is easy. What we don't understand is why certain small differences, from the absolutely trivial to the merely minor, yield such disparate responses, ranging from strongly attracting, to no interest, to being repellant.

Finally, these distinctions are not found in the features themselves but in how they are responded to. Of course, it is the sexual reactions of members of the opposite sex that are relevant, that are operant. For conscious beings, these include sexual feelings or emotions. But whether or not feelings are involved, the power of small differences, though undeniable, seems inexplicable. Why would minor, functionally meaningless differences in appearance and action evoke such dramatic differences in reactions and, where applicable, emotions? Viewed from the outside, there seems no greater reason for the varying sexual responses than the trivial differences themselves.

This carries us into murky waters. The phenomena of sexual attraction seem untethered, not from biology, to which they are inescapably fused, but from reason. If the only difference between an appealing incarnation and one that is unappealing is the appeal itself, then all we can do is state the obvious: that certain things, objects, and processes are more potent sexual attractants than others. And of course this is to say nothing about them beyond identifying their existence. What a terrible place to leave things. We simply do not have explanation enough. However daunting the task, science's job is to fathom such perplexities, to provide us with the rational basis for the different reactions. That is the subject of the next chapter.

Chapter 10

IN THE EYES OF THE BEHOLDER

Utility and Beauty in Sexual Attraction

As neere is Fancie to Beautie, as the pricke to the Rose, as the stalke to the rynde, as the earth to the roote.

John Lyly, in *Euphues and His England,* 1588

I love you without knowing how, or when, or from where.

Pablo Neruda, *100 Love Sonnets,* 2007

To sexually arouse a member of the opposite sex, the attracting feature must be observable. This of course means that the response depends on engaging the reacting organism's sensory mechanisms. For higher animals, attractants are *seen* as forms, shapes, or movements; *heard* as songs, calls, or merely talk; *touched* as hair or skin; or *smelled* as an odor. In what follows I will limit myself in great part to the first of these, to visual attractants, to what is seen. Whether or not they are the most broadly significant sexual attractants, they are certainly the most baffling.

Over the past half-century or so, it has become clear that in the mammalian brain even the simplest geometric forms, and certainly complex scenes and movements, are not directly perceived. They are only "seen" after they are analyzed. First, our brain asks, what is it—a chair, a giraffe, an elephant, or a truck? Then, is it alone, or is it with or next to something? And finally, is it moving or stationary and if it is moving, in what way? Seeing is not merely the direct physical conversion of a pattern of incoming photons to a mental image. Much is interpolated between the input and the perception. And it is through the interposed analysis that we come to understand objects as particular things, with particular properties.

Seeing a feature and determining that it is sexy is similar, except that in this case the analysis provides more than a physical imprint of some exemplified object. "She looks sexy" is not the same as thinking that "It looks like a tree" or even that "She looks like a woman." It is like seeing a rattlesnake and thinking not that it is a rattlesnake but that it is dangerous. We deem features sexy not because of what they are but because of what they mean to us. And yet, unlike a dangerous rattlesnake, and despite the fact that sexual attraction can be toxic, even lethal, we are pretty much in the dark about the reason for the attraction.

We know both as a matter of experience and scientific knowledge that a rattlesnake is dangerous because it can inject a poison that can kill us. But we cannot make a similar statement about a sexy attribute. We cannot say that this or that feature is sexually appealing because of some particular property, like a poison, that inheres to it. Though both observers and participants can readily and often with great specificity describe the reason for the attraction or lack thereof, they cannot explain its underlying basis.

There are two ways to deal with this inability. The first is to say that no explanation is possible, that the allure is not based on reason, whether conceptual (as a matter of logic) or mechanistic (as a matter of operational mechanism). I do not mean that the response to the attraction is poorly thought-out, but rather that it is not thought out at all. A rational judgment, however misguided, does not take place.

This is the proverbial, even the axiomatic, view of sexual attraction. Throughout recorded history, commentators of all stripes—philosophers, poets, novelists, playwrights, songwriters, and so forth—from the great and inspired to insipid hacks, writing of both the exalted and the depraved, have described the mystifying intricacies of human sexual attraction *and its amputation from reason.*

UTILITY

While such a determination appeals to the romantic mind, to science it is not merely uncomfortable; it is an anathema. If it were true, then we would have to conclude that sexual attraction is unique among natural phenomena. It would differ in a fundamental way from all others. Even psychosis has its own internal logic. Psychology and neuroscience hold that however misguided, however distorted, however remote from or only tortuously connected to reality, however crazy and groundless our thoughts and behaviors may be, they are the result of an internally consistent process of cause and effect. Even if the outcome is irrational, the thinking is coherent. The thought is the specified effect of a particular cause.

Even purposeless phenomena such as natural selection, unpredictable ones such as quantum effects, or random ones such as diffusion ultimately and unavoidably have a logical basis. To say that among natural phenomena sexual attraction is an exception, perhaps the sole exception not only beggars belief; it is deeply antirational. Why would nature set the features that provoke sex and reproduction, so critical to biological evolution, apart from logic and reason? However ignorant we may be of its rational basis, it must have one, and this is the second possibility, that the attraction is rational, not irrational. This does not mean that it is well thought-out or reasonable but simply that it is no less the consequence of a process of cause and effect than psychosis.

Procedurally, identifying rational grounds for sexual attraction seems simple enough. We know the effect, the attraction. All we have to do is discover its cause, the basis for the attraction. Not only that, but at least for visual attractants there are only two possibilities. A feature can be sexually attractive *because it is useful or because it is beautiful*. We can think of them as opposites—as in, it is beautiful but useless.

Given the "practical" nature of evolution, it seems both sensible and natural to favor utility. That is to say, certain features are sexual attractants because they signify the potential usefulness of a particular partner. The prospective partner will be a good, effective, advantageous mate. Perhaps he or she is vigorous, potent if male, fecund if female, and that as a consequence the mating will be successful and produce healthy offspring. In species where partnering involves parental responsibilities after intercourse or after birth, qualities such as the parent's maternal nature or the ability to provide and protect may come into play.

In humans we can think of the development of the female breast, the ampleness of her hips, the width of a male's shoulders, the size of his arm and leg muscles, and even rosy cheeks as features that are both attracting and utilitarian. But how much sense does this make? To begin with, there seem to be limits on the size of things for them to be attractive. A feature may be too big, breasts or muscles too large, hips or shoulders too broad. They may gain in attractiveness as a result of their size, but only up to a point. With additional increases, the attraction may be diminished, even lost. There can be too much of a good thing, as well as too little. How can this be explained?

Not only that, more importantly, an evaluation of utility based on appearance can be quite deceiving, as more than a fair share of men and women have discovered, having chosen infertile or otherwise sexually inadequate partners. Even rosy cheeks may not be a sign of health but of disease, a rash or fever. At any rate, since our choices can only be based on external characteristics, not actual fertility, our judgments can only be guesses. As guesses they may be either good or bad, but either way they can never be more than guesses.

In any event, however potentially valuable a determination of utility may be, it does not seem to take place for humans and perhaps in general. Outside of arranged marriage or similar calculating dispositions, humans' opinions about the sexual attractiveness of a prospective mate are much like the automatic reactions or reflexes of nonsentient animals. We are simply and directly attracted, even drawn, from afar. The desire is immediate, and the utilitarian suitability of the potential mate immaterial. As for some sort of unconscious calculation of utility, sexual attraction takes place in animals that lack cognitive function and hence lack the ability for such cunning.

Furthermore, as already touched upon, sex is not restricted to the most appealing among us. While in some species only favored males are chosen, and some males or females may be celibate by nature or circumstance, unavoidably impotent, barren, or otherwise unable to engage in sex, with the exception of extreme ugliness or severe functional impairment, whatever the species, however attractive or unattractive the individual, most sexually mature organisms reproduce.

Where an individual fits on a continuum of sexiness does not determine whether he or she has sex, nor does it govern the frequency of the sex or control how many offspring are produced. The less attractive procreate just as their more pulchritudinous brethren. To the degree that sexual attractiveness is a reflection of utility, the less useful reproduce just like the more useful in spite of their flaws.

And so the claim that attractiveness is a guide to utility is not borne out by the facts. This is reinforced by the fact that most things that make a mate suitable by utilitarian standards have no sexual valence whatsoever. Of course, our internal workings have none; the liver and stomach are just not sexy, however well constituted they may be. Even among features that our senses can perceive, not all useful traits are sexual attractants. Though it seems that humans can psychologically infuse any feature with sexual meaning, many that are quite useful—such as nostrils; nose hairs; the auditory canal; knuckles; premolars; sweating; evacuating; not to mention thinking; and, most importantly, eating and breathing—are not in the normal course of events sexually charged.

Conversely, as Darwin told us, many sexually attractive features have no obvious connection to the practical serviceability of a mate. Whether a face or bird feathers, functional adequacy is not correlated to sexual attractiveness. Homely faces and dull feathers do all the things faces and feathers do.

In the end, the conclusion seems unavoidable. Utility is not a satisfactory general basis to explain sexual attraction. Indeed, sometimes the idea of a utilitarian purpose for sexual attraction seems, well, nonsensical. For instance, a human parent anxious about their adolescent child's future, confronted by their strong attraction to an unsuitable, even a potentially destructive, mate, is not likely to see the child's choice as being the consequence some sort of conscious or unconscious understanding of an inherent utilitarian good in that person, whatever the child thinks.

BEAUTY

But having to reject utility is dreadful because it leaves only beauty as the source of sexual attraction, and this is hardly a promising place to look for the workings of reason. Over the course of human history, the question of beauty, including sexual beauty, has attracted the attention of some of humankind's greatest thinkers, but as the philosopher Roger Scruton points out in his book on the subject (*Beauty*), the concepts of truth and good have been far easier nuts to crack, and needless to say they are far from simple and far from resolved. Even when viewed with a well-studied aesthetic sensibility and with anthropological breadth, beauty is a slippery business.[1]

On top of this, the only chance we have of apprehending the aesthetic of sexual beauty in any sort of rigorous fashion is in our own species. It is only here that we have intimate knowledge and can speak with meaningful authority about it and its splendor. The simple fact is that we just cannot grasp how the peahen understands the peacock's feathers. We cannot know whether it sees them as beautiful or sexy or neither without talking to peahens.

Not only don't we know how peahens feel; we do not know whether sexual attraction in other species is based on the same aesthetic values as ours, or, for that matter, if it is based on any aesthetic of beauty whatsoever. All we can do, all that we can ever do when talking about beauty in nature is to express human sensibilities or extrapolate from them to other species, in other words, speculate. We can know the basis of the choice, but not the reason for it.

This is not to say that such speculations are worthless. Given the importance of sexual beauty in humans, there is good reason to think that other species, certainly those closely related to us, have similar perceptions. In fact, it is often hard to explain certain features and behaviors, especially in higher organisms, without assuming some sort of sexual aesthetic. That is, even though it is true that without talking to a peahen we cannot be sure what she is thinking, maybe she does see peacock feathers as beautiful in roughly the same way that we do.

In an attempt to put these uncertainties aside, I will limit the rest of this discussion to a consideration of sexual beauty in humans. In classical

Western art, taut breasts, ample hips, a narrow waist, graceful arms and legs, a well-turned ankle, and the gentle bulge of the belly that leads to Venus's triangle are depictions of the beauty of the female human form. While in the male we might point to broad shoulders, prominent pectoral muscles, a muscular rib cage, well-developed biceps and forearm muscles, and powerful leg and thigh muscles.

As for facial beauty, the handsome male face tends to be angular, while that of a beautiful female is gently curved. A prominent chin, high cheekbones, a straight brow, and a linear, relatively narrow nose are commonly perceived as beautiful in both sexes. In the female the upper lip forms a Cupid's bow and the lower lip is full. The eyes are deeply set with long lashes in the female and prominent eyebrows in the male.

None of this is meant to imply that there is a common understanding about what constitutes sexual beauty in humans. There are many cultural and racial differences, and even in the same group changes in the sexual aesthetic take place from time to time and period to period. Moreover, some women are attracted to heavily muscled men, while others are drawn to those with a slight frame, and while large breasts and a voluptuous body entice many men, others are attracted to thin female bodies, even those with prepubescent sexual features (note *Lolita*'s Humbert Humbert and today's models). And no doubt many males and females are attracted to both. Add that humans can, as said, mentally infuse almost any aspect of bodily appearance, from top to bottom, from feet to ears, with sexual content, we can see the enormous difficulty we face in trying to establish what is beautiful.

Yet we cannot relegate the perception of sexual beauty to some nether land as a subjective figment of the fertile human imagination. However difficult to understand, it has a biological basis. Whether it is hardwired or malleable, automatic (mechanical) or the result of some sort of intellectual, psychological, or cultural scrutiny, or all rolled into one, there is a biology of sexual beauty. However open to opinion and however varied and divergent that opinion may be, however subjective the sense perception may be, however obscure, sexual beauty has an "objective" biological basis.

We can see this in a variety of ways. Perhaps most striking is the fact that humans often extend great effort to present themselves to the opposite sex in a fashion that they think will be perceived as beautiful or handsome and that as such will be sexually appealing. This is most well developed in the female, and servicing the effort is an enormous industry in developed societies. We not only experience sexual beauty and see others experiencing it; we take action to amplify its expression, to make ourselves more appealing. However culturally dependent such adornments are, they give sexual beauty an observable third-person manifestation, an objective quality that at least in theory can be analyzed.

For instance, female clothing is often designed with sex appeal in mind in addition to and sometimes in spite of its utilitarian value. Though styles differ from time to time and culture to culture, clothing design often seeks to accentuate bodily features that are thought to be desirable and to hide those that are thought unappealing. And of course females have painted and decorated their faces for thousands of years to highlight its beauty as well as hide its imperfections. And humans produce all sorts of appurtenances that we attach to or otherwise place on our bodies, such as rings and tattoos that distort its form often for sexual aesthetic reasons. The focus on beauty can also be seen in it being *hidden*. For instance, in strict Muslim societies female beauty is thought to be so dangerous that clothing is designed *not* to accentuate bodily or facial beauty but to hide it (as well as less attractive features).

However obscure its basis and however varied its expression, sexual beauty is of great importance in the life and times of our species and probably many others. It is a potent determinant of sexual selection in humans and as such is vital to our evolution, an influential part of human biology, no less an aspect of nature than sunshine. Though it may seem ephemeral, something apart from nature, it is not. In this case it should be amenable to scientific analysis like the rest of biology. Yet, such an analysis is fraught with difficulty. Though in theory we should be able to determine why beauty inures to particular shapes, forms, colors, and movement, in practice the task is daunting and indeed may not be possible to carry out.

The *New Oxford American Dictionary* says that beauty is "the combination of qualities, such as shape, color, or form, that pleases the aesthetic senses, especially the sight." The question for scientific contemplation is: what is an "aesthetic sense"? Despite enormous differences among individual humans across time and culture, we all seem to possess a sense perception of sexual beauty. But it is one thing to say that such an aesthetic exists, and quite another to characterize it.

BEAUTY AND SCIENCE

In this light, before we can even attempt to analyze beauty's nature and pathways, we have to establish what we mean when we say that something or someone is beautiful. We need to define its rules and observational basis. What is beautiful and why?

Sadly for science (and perhaps wonderful otherwise), its *basis*, as opposed to its existence, is often shrouded and seems undecipherable. Beyond Euclid's geometry, Platonic ideals of balance, feng shui, and the like, and despite century upon century of aesthetic criticism and analysis, and however important in our lives, we remain almost as totally ignorant as the caveman in understanding the underlying biological basis of sexual beauty. However hard we try, we cannot define it in a way that takes us beyond the dictionary notion of a vague aesthetic faculty. Given its existence, we have no idea of what comprises it.

Accordingly, whether you find this heartrending or merely true, qualitative no less than quantitative generalizations about sexual beauty, about the beauty of certain shapes, colors, and movements seem beyond our comprehension. This is incredible since we not only make personal, subjective, and even intersubjective (among people) determinations of it every day, but our descriptions of what we think of as beautiful can be of great accuracy and reflection. But none of this is much help in attempting to fathom the underlying biological basis of sexual beauty.

Given this situation and human predilections, it should come as no surprise that some see this as a reflection not of our failure but of nature's incoherence. They would say that when you get right down to it, after all is said and done, beauty is just a subjective, personal, or cultural belief about which we are limited to mere description. Any attempt to uncover some sort of underlying rational or scientific basis for it is fruitless, even silly.

You can see why this assumption is appealing by asking any expert on beauty to explain why this or that object is beautiful. Why is one landscape painting beautiful, while another of the same scene, done with equal or even greater technical skill, is aesthetically displeasing? The expert will likely have a ready answer or explanation, and it may be convincing, even engaging, but in all likelihood at base it will be little more than an expression of his or her aesthetic preference or bias. As for sexual beauty, ask an artist to explain why small differences in the drawn line can change a figure from indifferent to one of remarkable beauty, or a face from homely to breathtaking, or vice versa? Or ask a clothing designer or cosmetologist to explain, *beyond describing what they do*, why it makes someone more attractive?

However precisely masters of the beautiful can describe beauty through their art, they are unable to provide us with an account of the underlying basis for what they create. I do not mean that they are ignorant of how a particular act makes something beautiful, or what they think of as beautiful, but why in an objective sense it *is* beautiful. Even Leonardo da Vinci failed in this quest. In the final analysis, despite it having a biological basis and hence some sort of underlying objective character, if there is anything about the material world that appears subjective and that lacks an objective basis, it is beauty.

And yet, science cannot acquiesce. It cannot accept failure. It cannot accept that sexual beauty has no specific meaning or series of meanings based in our brain and bodies that goes beyond personal opinion and cultural disposition. In spite of its obscurity, we cannot leave sexual beauty as some sort of gauzy subjective quality. It must have an objective basis that is amenable to scientific examination. *If only we could figure out how to do it*, if only we could figure out the terms of analysis. If Darwin's theory of sexual selection did nothing else, it made it clear that beyond any cultural aesthetic a scientific grasp of the nature of sexual beauty is critical to fully understand the mechanisms of reproduction and reproductive evolution.

SEXUAL HEAT

A hard-nosed reductionist might read all this and grumble that to focus on aesthetic notions to explain sexual attraction is ridiculous. There is after all a real, as opposed to an imagined, sex drive. When you look at *it*, all this hand-wringing disappears, and we can come to understand the nature of sexual attraction in scientific terms, reductionist terms at that. The first thing we must do is call it what it truly is, "sexual heat"!

For humans, to be sexually attracted to someone may actually make you feel hot. You may become flushed and begin sweating. In addition, your heart may race and your respiratory rate may increase. There is no beauty in any of this, just a physiological reaction to the attraction. We are simply and directly physically attracted to certain persons of the opposite sex. And when we are, there are no intervening thoughts. We just *feel* attracted and perhaps a bit terrified by the loss of control, even the loss of autonomy.

All of this is due to nothing more than hormones and neurotransmitters released in response to the appearance, movement, voice, scent, or touch of a prospective mate. In many animal species, the male is concupiscent, horny, and lascivious, constantly seeking sex, driven relentlessly and inescapably by his physiology, and he does so without taking into account the unique character of the potential partner. As for the female, she also seeks sex without contemplation; she is receptive when her sex hormones bid her. Add to this the hedonistic desire for sex in our own species, we find sex pleasurable, and nothing else need be imagined.

There is no contemplation, utilitarian or aesthetic, conscious or unconscious. Sexual attraction is merely a bodily function. When a partner is chosen, a best partner, or even a better partner, is not the point, just sex. Whether the choice is poor or first-rate is irrelevant. Pairing is simply a matter of opportunity and circumstance, the consequence of chemical and physical responses within us that are automatic and physiological, not cognitive.

Looked at this way, Darwin's theory of sexual selection is dead wrong. Even though partners are chosen and sex occurs, the choice is not informed by either utility or beauty, or anything else. There is no mental analysis, no aesthetic sense. If we view sex and reproduction in this sparse and unromantic way, it is no different in humans (and other higher animals) than in simple invertebrates, protozoa, and plants. If we rid ourselves of a sexual attraction based on vacuous notions like beauty or an intuited or inferred utility, biology is made one. We are no different than any other living creature.

This certainly simplifies matters, but the problem is that it dismisses what can't be dismissed. Sexual heat does not arise from thin air. It is a response to something. If that something is visual, as in the appearance of a potential partner, the response can only be due to the beauty or utility of what we have set our eyes on. When you get down to it, sexual heat or, more accurately, sexual desire is generated either by the utility or beauty (or both) of what we see. Talking of sexual heat does not get us off the hook. We have not unburdened ourselves. Why does a particular appearance give rise to feelings of desire?

Though science cannot let it stand unexplained, if we reject utility and can't understand beauty, how do we make sexual beauty comprehensible? Even if stolid utility is unable to explain sexual beauty and the idea of beauty itself is incurably murky, sexual beauty still produces desire, desire produces sexual selection, and that in turn leads to sex and reproduction. If as Darwin thought natural selection has no place for beauty, given the existence of sexual attraction, some other kind of an explanation was needed, and hence his theory of sexual selection.

IN ITS OWN RIGHT

Beauty runs in families. Its traits are inherited. They also appear to have evolved not for some utilitarian demand of natural selection and survival but to engender reproduction. In both its makeup and perception, this is why it exists.

This is at the same time self-evident and revelatory. It is self-evident that sexual beauty drives sexual selection, and that this in turn provokes sex and leads to reproduction and life's continuance. It is revelatory because whatever is deemed beautiful is useful because of its beauty, not the other way around. What we find is the usefulness of beauty, not the beauty of usefulness. It is not utility ergo beauty, but beauty ergo utility.

This realization leads us further. What distinguishes the more from the less beautiful? What agency is responsible for its emergence? Surely it is not natural selection with its exclusive focus on survival. We may see beauty in natural selection's world, but natural selection does not. Not only is there a need for some agency other than natural selection to account for traits of attraction and beauty, as we shall see, whatever it is, it is needed to account for life's reproductive features. If not natural selection, then what was responsible for their evolution, by what means did these traits evolve?

Organisms face two barriers to reproduction—natural selection and sexual selection. For reproduction to take place, and hence for evolution to take place, both have to be affirming. As said, one has to survive to maturity (natural selection) *and* have sex (sexual selection) to create the next generation. Only then can evolution occur. The former, natural selection, is responsible for the evolution of all the properties of life except for those of reproduction. The latter, the act of sexual selection, including the attributes of beauty are the result of something else, a *reproductive selection* of some sort.

As we shall see, this reproductive selection applies not only to sexual selection but also quite generally to the evolution of the mechanisms and processes of reproduction. And yet, if the reproductive mechanisms of life are not the result of natural selection and its survival, then what is it about? What is reproductive selection based on? What purpose does reproduction serve? Indeed, does it have a purpose? To what end does it exist?

These questions may seem especially silly because the answer appears to be patently obvious. Of course reproduction and along with it sexual beauty and choosing a mate has a purpose. That purpose is to produce new living things. But as self-evident as this may seem, as we shall see, the answer is unsatisfactory. In section V, "Purpose and Reproduction," we will explain why and search for a satisfactory answer, as well as for a selective process that fits reproduction. But before we can do this, we have to consider the idea of purpose in biology in general terms. What does it mean to say that reproduction or any biological process has a purpose?

SECTION IV

PURPOSE

Chapter 11

A GREAT CONTRADICTION

Teleology Rears Its Ugly Head

Cat: Where are you going?
Alice: Which way should I go?
Cat: That depends on where you are going.
Alice: I don't know.
Cat: Then it doesn't matter which way you go.

Lewis Carroll, *Alice's Adventures in Wonderland*, 1865

This brings us to a great contradiction. There are two propositions about the nature and genesis of the reproductive mechanisms of life, each seemingly true in its own right that gainsays the other. They are as follows:

- The mechanisms of reproduction and the means of their evolution have the purpose of ensuring the future of life.
- The mechanisms of life, including those of reproduction, and the means of their evolution are mechanical and therefore without purpose, and as such are not about ensuring the future of life or anything else.

The first proposition tells us that reproduction expresses an *inherent* concern for life's future. The second, in line with science's understanding of the workings of the universe, tells us that the mechanisms of life and its evolution, including those of reproduction, take place *mechanically*, without intention and without interest in life's fate.

There is little doubt that the consensus of modern science falls on the side of the second proposition. Putting human desires and intentions aside, all events in the material world are thought to be without purpose. In regard to life's evolution, the odds of organisms surviving challenging situations in the moment or the long run, within or across generations, is determined by the causally indifferent circumstances of Darwin's natural selection. This lack of interest applies no less to reproduction and the evolution of its mechanisms than to life's somatic features.

Yet the first proposition seems self-evidently true, even incontrovertible, and no more than a simple statement of what reproduction is. There is, however, a way to avoid having to choose between two propositions, each of which seems true in its own right. We can maintain that both are true, that they are not contradictory but paradoxical. We can say that in a display of cosmic irony, a physical agency with no interest in life or its future introduced purpose de novo. In what follows, we will consider this possibility for the reproductive features of life. But before we do, we will assess it in regard to life's somatic traits and its natural selection. Do the somatic traits of life serve a purpose introduced by the purposeless natural selection?

THE FOUR CAUSES

We start not with the most modern of modern molecular biology but more than two thousand years ago with the great philosopher Aristotle. He said that nature in all its variety could be accounted for in terms of just four understandings, translated usefully, though somewhat inaccurately, as "causes":

1. *Material.* Aristotle talked of material causes as the brass and silver from which vessels are formed. In a modern scientific context we might talk of the underlying nature of matter, of molecules and atoms, of subatomic particles, even of energy.
2. *Formal.* A formal cause is the "form or pattern" of the object, its "essential formula."
3. *Efficient.* According to Aristotle, an efficient cause is the "beginning

of change or rest." The father, he explained, is the efficient cause of the child.

4. *Final.* Again Aristotle offers an example. If we walk for health, then health is the final or ultimate cause of our walking, of our movement. This is its purpose and the reason for its occurrence.[1]

Putting the four causes together in an example, we can say that the material cause of a chair is wood (or, if you wish, what wood is made of), its formal cause is its form or pattern, its efficient cause is how it is produced, and its final cause is that it is meant for humans to sit on. As for the place of Aristotle's causes in modern science, the British philosopher and scientist Francis Bacon set the stage three centuries ago. Science, he said, is about Aristotle's first and third causes, material and efficient causes. In today's language, it is about statics (material) and dynamics (efficient).[2]

He dismissed formal and final causes, placing them outside of science and making them the subject of controversy ever since. In the first place, it was argued that formal and final causes are really the same thing. They are what make a thing what it is. The formal cause of a chair is its form or pattern, the formula for the chair. Is this not the same as its final cause, what it is intended to be? It is intended to be what its formula specifies.

Though despite their abstract nature the existence of formal causes is accepted by many today, human inventions aside, the proposition that there are final causes has not. It has been called Aristotle's greatest error.[3] Things merely come to be; they develop or evolve (efficient) in accordance with some form or pattern of nature (formal) from material antecedents (material). There is no other reason for being or for actions. There are no ultimate purposes (final), such as good health, however much we might desire it. Things, including living things, simply become, and having become, just are.

TELEOLOGY

In keeping with this view, modern reductionist and materialist science is often understood to deny the presence of final causes and consequently to deny the presence of purpose in all of creation, in life just as in the stars and planets. Not only doesn't science welcome purpose into life; it excludes it.

As a graduate student in physiology at the University of Pennsylvania more years ago than I care to remember, I was told to avoid something called teleology in my writing and thinking.[4] Teleology was an expression of Aristotle's final cause, *telos*, and to invoke it in reference to anything biological, any structure, process, or action, was to say that it existed for a particular end, that it had a purpose, and this was unacceptable to my professors.

What made teleological thinking so objectionable was that it implied, or so it was thought, the existence of a deity or designer that imposed purpose on the object. Such causes were rejected either because it was understood that they were not part of the physical or material world and that world comprised the whole of existence, or, more benignly, because causes external to the physical world were beyond science's purview.

As students, we learned that as a matter of science it was not acceptable to say that the heart had the *purpose* of pumping blood through the circulatory system, though this was of course what it did. Rather, we were told to say that it *functions* in this capacity though this was not its objective. To imbue the heart with a purpose was a logical error. It was to say that nature *intended* it to pump blood, when it did no such thing. This was teleological thinking and it was unacceptable.

DARWIN'S THEORY IN A PURPOSELESS UNIVERSE

Remember that Darwin's theory has three critical elements, in order of occurrence: mutations, natural selection, and reproduction. First mutations, random alterations in our genes, in our genetic or inherited makeup take place. They are understood today to be changes in the chemical structure of the DNA molecule. If a mutation alters an organism's character in some meaningful way, then (second) natural selection acts on the organism in reference to this feature, and (third) if thus constituted it survives environmental challenge, reproduction occurs and the mutation is inherited. This process of mutation, selection, and reproduction is repeated over and over again, generation upon generation, down through the ages with new traits being exposed to environmental challenge and old ones being confronted by new challenges. The result is evolution.

In Aristotelian terms, our embodiment is evolution's material basis, while reproduction is its efficient cause. The genetic code embedded in DNA provides a common (essential or unique) form or pattern, or general formula for the enormously diverse products of evolution. It is life's formal cause. And finally, natural selection, though critical, is not a cause at all. It is a choice, nature's choice. Accordingly, life evolved based on our embodiment (material), according to a genetic formula (formal), by a process of reproduction (efficient), with natural selection guiding events.[5]

Notably absent are final causes and their purposes. In a purposeless universe, they are neither present nor needed. The magic of Darwin's theory was that it offered an account of life's evolution that had no need for them. All that was required were material, formal, and efficient causes, and the aimless agency of natural selection. Final causes with their purposes and intentions were not necessary. Evolution was mechanical.

PROGRESS THROUGH NEGATIVITY

In this view, even though evolution was progressive and creative, oddly, its agency, natural selection, was ceaselessly negative. Selection occurred by the selective *elimination* of particular varieties and species. The unfit were eliminated, leaving the unaffected or less affected behind to evolve. Evolution was a process of default or exclusion, not, as often imagined, one of affirmation in which the fit survive. Though it is true that the fit do survive, life evolved through the elimination of the less fit.

The affirmative aspects of evolution can be attributed to mutations (as well as to gene sorting during reproduction), not natural selection. A small fraction of random mutations enable the expressing organisms to survive particular environmental challenges that those that lack the mutation cannot survive or survive as well. As such, the accumulation of favorable mutations over eons served as the material source of the adaptive properties of life. They account for everything about us, from our predatory behavior to the workings of the circulatory system.

Yet, however useful in their own right, mutations do not give rise to the adaptive map of life. They may improve a particular property and as a result make an organism more robust. But, given their random nature, a further or different mutation to the same gene would in the normal course of events, and as chance would have it, inevitably annul the effect or even produce a contradictory one. Mutations tend to cancel each other out. They do not lead to an enduring accumulation of positive, that is, adaptive features. Indeed, commonly they are neutral or harmful and only rarely helpful. Mutations reverse valuable changes just as they make them. As such, they produce change, *not evolution*. Over time, without the help or guidance of natural selection, as mutations continue to take place randomly, traits that are adaptive would invariably be lost and replaced by others that are neutral or harmful merely as a matter of chance.

But natural selection does take place, and evolution is the consequence. What it does is introduce a bias. Natural selection preferentially eliminates creatures (and their mutations) that lack the ability to overcome particular hostile circumstances. As a result of this selection, in some cases leading to death, in others to less effective propagation, unaffected or less-affected organisms are advantaged by exclusion. In this way, without purpose or intention, natural selection favors the survivors simply by not viewing all mutations with, so to speak, equal disdain.

And so natural selection "produces" destruction. Its effect is invariably negative. Cold temperature or an attack by a predator may kill you, but it can never help you. You may find ways around it, but its impact, when it has one, is always negative. Natural selection either produces death or predisposes an organism to it, and may even thereby doom a species to extinction.

Whether devastating or hardly noticeable, in the end death and extinction are the only possible outcomes of natural selection. Though often thought of as being progressive and productive, when effective natural selection is quite the opposite. It slouches toward the end of life and general disorder. It increases the randomness of the system. It increases entropy. Manufacturing nothing, ultimately natural selection is annihilating.

Leaving reproduction aside, life in all its variety and magnificence seems to have evolved in this counterintuitive fashion. It evolved as the result of a combination of random and entropic events, random due to mutations and entropic due to natural selection. Together they are not only without purpose; they are negative, affirmations of the second law of thermodynamics and entropy's tendency to increase.

If the evolution of life were based on mutations and natural selection alone, then like everything else in the universe, life and its adaptations would lack purpose and final causes, extending the aimless cosmos to life. Our multitudinous, multifaceted anatomical and physiological features, the adaptations that protect us from all sorts of environmental threats, would have evolved and would function in the absence of purpose. They would be the product of random mutations that just happen to spare particular species or reproducing groups from the entropic action of natural selection (for the time being). They evolved by surviving or eluding the destructive effects of natural selection. As luck would have it, the spared groups are given the opportunity for further modification and development through the addition of other helpful mutations. They are thus given the chance to evolve unless and until a changed environment, including the appearance of other species, makes it more vulnerable and leads to its extinction.

THE END OF TELEOLOGY

According to this prescription and the argument against teleology, and in line with modern materialist science more broadly, the anatomical and physiological adaptations that comprise our (somatic) being and that protect us from all sorts of environmental threats are understood to do so without intention, just like the heart pumping blood. We say that their actions and the work they perform lack *purpose*, as does their coming into being and their evolution. Yes, adaptations are salutary and help in our struggle to survive, and we endure the challenges of natural selection because they are, but our survival is not their aim or goal. They have none. What they do, they do without intention, without purpose. If a specific feature helps us, so much the better, if it does not, so much the worse, but whether it helps or not, the outcome, whether we survive or succumb, whether we are advantaged or disadvantaged, has nothing to do with what is intended. Adaptations intend nothing.

And so, living things, comprised of purposeless elements, are purposeless themselves, no different from rainfall or a flowing stream. Whatever they do to live or live to do, whether breathing, circulating blood, eating, or even thinking, lacks purpose, as does life itself. Obviously, this conclusion is enormously consequential. No matter how widely held, and no matter how confident one may be in the belief, before setting such a momentous conviction in stone, we should be as certain as we can that the belief is justified. To make this determination we need to look more closely at Aristotle's final cause and its relation to life. Is there any provenance, any status at all for final causes in living things? Is life truly without purpose, or does it somehow evince it?

Chapter 12

PURPOSE

Teleology Cannot Be Denied

> *A purpose, an intention, a design, strikes everywhere even the careless, the most stupid thinker.*
>
> David Hume, *Dialogues Concerning Natural Religion*, 1779

This takes us back to my graduate school days again. It turned out that despite our importuning professors, we students found that even if we scrubbed our writing of suggestive verbiage, it was no easy task to avoid teleological thinking in biology. The reason was simple enough. We were naturally drawn to the conclusion that if a structure pumped blood, then in all likelihood this was its purpose. And if this was true, then it must have evolved with this in mind. However faulty our understanding, to imagine that things that appeared to serve particular purposes, even patently obvious ones, served no purpose at all, flew in the face of, if not philosophy or logic, then common sense.

Why, for example, we might naively ask, if the cosmos, with its living things and their subsumed mechanisms, is purposeless, do we even have words like *purpose*, *intention*, *desire*, *goal*, *end*, and so forth pointlessly littering our vocabulary? Were we driven to conceive of the inconceivable by some perverse creative force, or perhaps for some inexplicable reason we are bent on deceiving ourselves about our true nature? But either way, what would motivate us to do so? On what basis would such a creative force or the desire for self-deception exist without the *purpose* of creation or *intention* of self-deception?

And if words like *intention* and *purpose* are no more than meaningless jabberwocky, then why do they seem to so accurately describe the events to which they refer? In any event, given the existence and use of such language, is it valuable to ask whether it is an expression of reality, of the way things are, rather than a creative curiosity or a conscious or unconscious psychological deception?

PURPOSE IN HUMANS

Graduate student misconceptions and naïveté aside, there can be no serious doubt about the answer to this question. The words are neither curiosities nor self-deception but rather describe reality. For humans, as we go about our lives, we have purposes and intentions of all sorts, ranging from the trivial to the life altering. Our almost every action is undertaken with a purpose in mind. It is with purpose that we brush our teeth, bathe, get dressed, prepare and eat food, wash dishes, lift objects, calculate numbers, make telephone calls, talk to each other, do things of interest with each other, go to work, make money, spend money, shop, run errands, watch TV, listen to music, play music, read, write, draw, surf the net, exercise, play a game, get married, decide to have children, raise children, engage in dangerous pursuits, refrain from them, seek to prolong life, determine to end it, and otherwise make all manner and means of choices to carry out as many activities.

Not only do we act to achieve particular purposes, we seek purpose in our lives. We have an enduring need to believe that however nugatory, what we do, how we live our lives serves some meaningful purpose beyond merely breathing in and out, eating and excreting. And we have this need regardless of our status or what we think of as a meaningful purpose. Parents worry endlessly about their grown children finding it. Depressed people as well as old people, particularly when they become seriously ill, often lose it. And in its absence, all hope seems lost and life hardly worth living.

But even if we could somehow convince ourselves that all these activities as well as many others that I could catalog lack purpose, we still would not be able to deny purpose a central place in human life. It is there and it is central. This is because humans design and build material structures, like chairs and machines, as well as immaterial ones, such as poetry and computer programs. These artifacts, whether estimable or worthless, serve the purposes intended by their inventors. Humans are their source; we are their creators, their final cause.

And so, however you look at it, we cannot exclude purpose from our lives. It seems a near scientific and logical certainty that humans have purposes, and if they exist for us, given that we are natural beings, how can we exclude their presence in nature otherwise? Even if we could find a way to exclude purpose from life's ordinary activities, seeing them as purely mechanical, there is no doubt that we manufacture things for a reason, for a purpose, and purposes have final causes. They exist in nature even if we insist that it is only for humans and their artifacts.

HOMO SAPIENS ALONE

No matter how at ease modern man may be with the notion of a disinterested cosmos (Darwin's theory included), there is purpose in the cosmos for no other reason than we, however insignificant and inconsequential, are part of it. Aristotle's greatest error was not an error at all. He understood perfectly well that even if we could exclude purpose from every other aspect of our being, in conceiving of and building things, humans create artifacts with purposes in mind, purposes for which we, godlike, appear to be the final cause.[1]

Before science's ascendance and before Darwin, the special place of humans in the world was the standard belief. Caterpillars may metamorphose into butterflies, and birds can fly, but human cognition seemed to be something apart, with intentions and purposes that went beyond nonhuman life. This was not mere solipsism, a sense of human self-importance, though it certainly was that. Humans seemed to differ from most other living things in a fundamental way. Human action seemed to be undertaken with purposes and intentions that other species neither had nor could imagine. Only *Homo sapiens* allowed for the creation of such magnificent things as machines and poetry.

This conclusion was simple common sense, but at the same time it was profound. It was profound because, as Aristotle explained, purposes have final causes, and final causes are elemental. That is to say, the capacity for purpose wherever it exists is a basic attribute or quality of existence. Its special presence in the human mind made us unique creatures, fundamentally different from the rest of the material world. For all our flaws, we stood astride and above the rest of life, not as gods but as creatures made in God's image.

MORTALITY

This high view of humanity faced one enormous problem. Not only did our physical incarnation bear a remarkable resemblance to many other animals; like them we were mortal. Our bodies were just as vulnerable as theirs to the predations of life. Some two hundred years before Darwin, René Descartes sought to resolve the incongruity between man the special being and man the ordinary animal. He simply set the human mind filled with its ideas and purposes apart from our mechanical body. It and our immortal soul were separated from the corporeal aspects of human life. Our physical and chemical embodiment, including what the heart is and does, was at one with the rest of the world of living things, and for that matter the rest of the material universe, mechanical and without purpose. But the mind rose above the mundane and the mortal.

In a modern context we might say that the human nervous system, or at least certain parts of it (notably the cerebral cortex and some aspects of the limbic system and cerebellum) are not only imbued with special capabilities such as our exceptional intelligence and motor dexterity, but also with purposes and even final causes, with fundamental properties that are not found in other species or even in other parts of our own bodies.

AGAINST HUMAN EXCEPTIONALISM

Stripped of its metaphysical context and godly design, the problem with Descartes's dichotomy was that it presupposed the existence of a basic law of nature, a law of purpose and final causation that applied only to humans and their brains. But there is nothing in biology or the physical world otherwise to suggest the existence of such an exclusive law. The laws of mathematics, physics, and chemistry, and the rules of logic more generally, apply equally and without exception to all living things, plants, animals, simple, complex, high, or low.

Descartes's separation of life into the special (eternal and lofty human mental activity) and the ordinary (the ephemeral and coarse mechanisms of the body) seemed to fly not only in the face of reality but also of modern biology. To claim that purpose and final causes, inherent or arising from God, are only associated with the human mind is to ignore the presence of relatively well-developed brains and nervous systems in other animal species. According to the modern view, as extraordinary as we humans are, we are in all respects just one animal species among many. Physically and chemically, in both body and mind, we are at one with both our close and remote relatives. Our genetics are based on the same code; we carry out the same kinds of chemical reactions; have all sorts of anatomical and physiological attributes in common, including the organization and mode of action of the brain. While every species has its own peculiarities, its own special features, life at the most basic level is the same phenomenon for all living creatures. Humans are not beasts of a different sort. Not even our minds can be separated from the unity and commonality of life.

Add to this that modern evolutionary biology has steadfastly refused to place humans at the pinnacle of life's tree, no less as a species apart. To it we are merely one recently evolved species, a fragile sprig in the history of life. In the greater scheme of things, we are probably of less significance than cockroaches, and certainly no different in kind. Human exceptionalism has no currency in modern science's view of life.

FINAL CAUSES IN LIFE

The fact that the mental activity of humans cannot be held to be something fundamentally different from the rest of life presents an important choice about purposes and final causes. Either they must be excluded from life in its entirety, and this includes our mental activity, or they must be included in all life, in everything biological, from photosynthesizing plants to the heart pump. Either all is without purpose, mechanical, or all is with purpose and hence has final causes.

Looked at this way we are no longer at odds with other life-forms. We are not odd man or woman out. If life has purposes and final causes, then purposeful human mental activity is not something apart from the rest of biology but is an expression of it. In this view, purpose and its underlying final causes must be a common, even a general, property of life and its evolution. And if purposes exist broadly in the world of living things, then life not only has material, formal, and efficient causes, but also final causes, just as Aristotle said. This relieves us of the burdens of an existential nihilism that denies purpose or alternatively of claims of a human exceptionalism that is at odds with both common sense and modern biology. In the next chapter we consider the possibility that evolution is purpose driven and how such a proposition would affect Darwin's seemingly purpose-free theory.

Chapter 13

DARWIN'S TELEOLOGY OF SURVIVAL

A Scientific Basis for Purpose in Life

Darwin's great service . . . (is) that instead of Morphology versus Teleology, we shall have Morphology wedded to Teleology.

Asa Gray, *Nature*, 1874

This brings us to Darwin, the teleologist. How odd, Darwin a teleologist despite the great significance of his theory being its mechanical nature, as a purpose-free, teleological-free design and God-free account of life's evolution. As explained, the crux of the theory was that it offered an explanation for how life evolved without intention, plan, or purpose, and yet it seems that its inventor came to believe that biological processes had purposes.

Darwin did not say much about his teleological inclinations, though he referred to final causes a number of times in his notes and letters and three times in the *Origin*.[1] We can see it in this appreciative reply to Asa Gray, the author of the epigraph above, connecting his theory to teleology: "What you say about Teleology pleases me especially and I do not think any one else has ever noticed the point."[2]

Did Darwin's teleological leanings merely signify his confusion? Or perhaps he was attempting to reconcile the irreconcilable, seeing his theory on the one hand as materialistic and on the other as teleological? Given his limited comments on the subject one cannot be sure, but in all likelihood he saw it both ways at the same time and found no inconsistency in this. We can understand why by asking what kind of teleology Darwin was likely to have ascribed to.

There are two kinds, God-based and natural. The God-based variety, often credited to Plato, imagines a contemplative Demiurge or Godhead, a thinking God aware and possessed of superhuman, though in some respects humanlike thought processes that imbue us with purpose, just as we do our constructions. Whether the formless eternal God of Abraham and Moses or the humanlike embodiments of the Greek gods, gods both imagine purposes and *bring their means into being*.

In the natural form of teleology, usually attributed to Plato's student Aristotle, God's role was one of conception and oversight, while nature was the instrument of purpose. Purpose was born of nature, not God. It arose not from the hand of the designer but as the result of natural processes. In this sense Aristotle's teleology and modern materialism, including Darwinism, are perfectly compatible. As for Darwin himself, it is hard to imagine that he held a Platonic view of teleology having unabashedly rejected Godly design in his theory of evolution. Though his teleological instincts seem to have been little more than a sense of things, however unformed, in all likelihood he ascribed to an Aristotelian perspective.

IS DARWIN'S THEORY TELEOLOGICAL?

Though Darwin's personal thoughts about teleology are of interest, the role of teleology in his *theory* is more than that. It is of great consequence. Though the theory is invariably thought of solely in materialist terms, it can be understood as being both teleological and materialistic, not one or the other, but both at the same time. This is troubling since the great importance of Darwin's theory lay in its materialistic, *nonteleological* explanation for life's evolution.

However little he said about it, from his comment to Asa Gray it seems that Darwin was aware of the duality. His insouciance may have been a political calculation. At the time, many scientists and laypersons would have viewed the inclusion of teleology in the theory as proof of its failure. It would testify to the fact that it did not provide a purely mechanical explanation for evolution. Since this was in great part its raison d'être, as well as its claim to fame, raising the theory's teleological shadow side might not merely have forestalled its triumph; it might have sealed its doom.

Still, whatever the threat, the theory can be understood in teleological as well as materialistic terms. From a materialistic viewpoint, the emergence of adaptations was the result of undirected occurrences. As explained, the process begins with random mutations, then natural selection acts on the organisms that express them, and successful mutations are memorialized by their reproduction. As a result, adaptations were created. Reproduction aside, adaptations were the purposeless products of purposeless events.

With a teleological mindset, the same events have a different cast. Though the operating agency is still natural selection and it remains as purposeless as ever, *in their emergence adaptations gain purpose.* Their purpose is both specific and general. It is specific to each particular adaptation, to what it does, such as pump blood. While at the same time each adaptation reflects the universal or general purpose of life's evolution.

What do I mean by evolution's "general purpose"? Darwin's theory not only tells us what that purpose is but is based on it. The evolution of life took place with a single overarching and all-embracing purpose—*survival.* For each and every situation, whether on the attack or under attack, whether vulnerable or needy, whatever the particular adaptive means, living things respond with the steadfast intention of surviving their circumstances. Each adaptation in addition to its specific purpose plays a part in fulfilling this grand objective of life.

The argument for a teleological view of life's evolution rests on a claim and an expectation. The claim is that by definition, for living things survival is a purpose, not a purposeless fact of life.[3] The expectation is that if this is true, we should be able to identify the source of that purpose. We should be able to identify its final cause, and hence be able to answer the question: What is it that causes an object or action to serve the purpose or end of survival? If the claim of purpose is warranted, and if it is possible to identify its final cause, then the teleological argument is successfully made.

PURPOSE ABIDING

There are only two places to look for purpose in the physical world. Either it can be found within or immanent to the referenced object, or it is external to it, that is, in its environment. As laid out in the last chapter, for humans the case for immanence is convincing, though as we shall see, it is not unmistakable. In this view, we serve as our own source of purpose. It arises from within us, from within our material being.

Whether significant or trivial, large or small, we express purposes in great abundance every day of our lives, in almost every waking moment of our existence. Almost everything we do from the time we get up in the morning to the time we go to bed at night is the result of our resolve to do so. Given this, the idea that life's evolution is simply the consequence of random mutations and entropic natural selection (followed by reproduction) is either insufficient or false. If it is either, and if living things, humans first among them, have purposes, then the conception of purpose must somehow fit into Darwin's mechanical theory, or the theory must be rejected.

It is simple enough to insert purpose into the theory. We just have to say, as I have, that survival is the general *purpose* of evolution and that each adaptive property serves this grand purpose in its own way. The problem is that this contradicts the theory's central premise. Indeed, this, rather than political calculation, may have been why Darwin said so little about his theory's teleological side. Having rejected purpose in its formulation how could the theory then survive its inclusion? After all, this is what made natural selection such an appealing concept. According to its harsh view, survival or the inability to survive is the result of surrounding influences, inanimate or biological, that are no different in kind from those faced by an eroding rock, and there is no worldly purpose in that.

It was the idea of geologist Charles Lyell, Darwin's friend, about geological evolution, uniformitarianism, that a slow change in the form of our planet's surface accounted for its evolution, that served as the wellspring for Darwin's ideas about life's evolution. Extrapolating, he imagined that life evolved in much the same way. It matters not to the rock or, for that matter, to the universe whether or not it is eroded. Nor does the formed valley, raised mountain, or altered shoreline indicate any purpose to the elements that comprise or form them, or to the cosmos more broadly. This, Darwin thought, as have many who followed, some with extraordinary fervor, was also the case for life.

Darwin understood that the analogy to geology was far from perfect and that living things were different in many ways from rocks. Nonetheless he thought it was apt and viewed life's evolution as being no more purposeful than the fate of a rock. But whatever Darwin believed (and his opinion changed over time) and however comfortable many are with this view today, there seems little doubt that survival is the central, yes, *purpose* of life. However purposeless the evolution of rocks, mountains, and the like may be, it is our undeniable and unshakable purpose to survive the challenges presented to us by the world we inhabit. That world may not care a jot about our continued existence, but we certainly do. Darwin's theory tells us that *our purpose is to survive*.

The poet Dylan Thomas pleaded, "Do not go gentle into that good night . . ." and we do not. Living things struggle mightily to survive even the most adverse and seemingly hopeless conditions. Think about one creature caught in the jaws of another, fighting desperately to extricate itself. Consider animals of all sorts, trapped by circumstance, an ant, a beetle, a bird, a small fish, a creature in the web of a spider, the orifice of an anemone, stuck between rocks, in mud, clinging to an edge, freezing, overheating, having lost the use of a limb, wounded, blinded, and on and on, struggling with every ounce of its being, every means at its disposal to survive its terrible situation.

And things are no different for a predator. He or she is hungry and needs a material source of energy to maintain life and to this end seeks food. In this predators are no less engaged in the pursuit of survival than those they act against. Even in the simplest species, the need for sustenance does not give rise to random chemical or physical events but to action that not only serves the *purpose* of obtaining nourishment; it embodies that purpose.

Even plants and microorganisms deploy all sorts of defenses to stave off destruction. And the newly born learn quickly and over time how to survive their circumstances; they learn the tricks of the trade, and that trade is survival. Add that a potential victim may hide, run, or even counterattack, and it should be obvious that whatever the character of its efforts, life struggles for survival and does so until the bitter end, that is, until the acceptance of death is the only choice. Unlike rocks that, however resistant to destruction, passively comply with the forces imposed on them, living things are active, tenacious, and give up only when all hope is lost.

If the world is without purpose, why struggle? Why fight so furiously to stay alive? What difference does it make in a purposeless and indifferent world? Why should we care, on what basis? Why don't living things simply go gently into the good night, why do they resist with such determination? Why struggle with such grit? Why not just submit? Rocks erode, life ends. It takes a very determined (and I might add purpose-laden) mind *not* to see that an organism in extremis is expressing the most fundamental purpose possible—the *desire* to survive.

AGAINST PURPOSE

But you might wonder if life in all its variety and magnificence did indeed evolve as the result of a combination of random and entropic forces, mutations and natural selection respectively, how would purposeful action arise, for there is purpose in neither? If, like everything else in the universe, life's evolution took place without purpose or intent, how could living things have come to embody it? How can adaptations have purposes if the mechanisms of evolution that produce them do not?

Leveling life, equating it with the inanimate realm, is based on two propositions: one of parsimony, the other an analogy. As for parsimony, we say that it is simpler not to impute purpose than to impute it, and that as such the claim of an absence of purpose is to be preferred. As explained, Occam's razor, the simplicity principle of science, states that the theory that requires fewer propositions is most desirable. The presence of purpose, with its final causes, *adds* to Aristotle's other causes and hence fails the test of parsimony.

The analogy is based on two related inferences. The first comes from an enormous body of reductionist research on life's chemistry. According to this view, because the chemistry of biology, including DNA chemistry, is, in its own right, that is, isolated from living things, without purpose, it follows that what it gives rise to in situ is similarly purposeless. If purpose is not found in the isolated part, then it does not exist for the whole.

The second inference has it the other way round. It goes from the general to the specific, from the broader system to a particular aspect of it. It states that if the broader system evinces no purpose, then neither does any particular aspect of it. The broader system I am referring to is the broadest of all, the cosmos. If the cosmos is aimless, if it exists without purpose, then how can any part of it display purpose?

And so, the presence of purpose in living things can be dismissed in two ways. Its absence in them can be said to reflect its absence in life's chemicals and their reactions, in the intimate elements of life. It can also be discounted because the world beyond living things, up to and including the stars and planets, is without purpose. Elsewhere I have called the first of these "strong micro-reductionism"[4] because it claims that the whole can be completely understood (strong) as the sum of its parts (micro). In this case, if the parts are without purpose, then so is the whole.

GALVANI AND PENFIELD

In the late 1700s the Italian physician and physicist Luigi Galvani provided important experimental support for a mechanical view of life. Galvani applied an electric current to the (sciatic) nerve that innervated the leg muscles of an incapacitated frog. The frog's leg muscles contracted, twitched. This occurred without purpose for the frog lying flat on its back on the dissecting table. The frog certainly did not intend to contract its muscles.

In the late 1940s, Wilder Penfield and Herbert Jasper at McGill University in Canada performed parallel but far more extensive and expansive studies with the same result. Rather than stimulating a peripheral nerve, such as the sciatic nerve to a limb, they stimulated certain areas of the brain (certain areas of the cerebral cortex) in patients before they underwent surgical treatment for epilepsy. They found that when a specific area of the cortex was stimulated with an electric current, it produced the contraction of particular muscles. By matching the location of stimulation to the location of the responding muscles, they were able to map the *motor cortex* or motor field of the brain, muscle by muscle, muscle group by muscle group, in the process producing a homunculus. As with the frog's leg muscles these contractions, from fingers to toes, served no purpose for the patients. Penfield and Jasper had imposed the movements on them.

In both studies, muscles contracted as simple mechanical responses to nerve stimulation. The contractions were purposeless except for the intentions of Galvani and Penfield and Jasper. Muscle contraction and movement could be understood in purely physical terms. This conclusion was supported by a variety of disease states or disorders, such as epilepsy, in which muscles contract in a haphazard and uncoordinated fashion, without either purpose or utility.

PURPOSE CANNOT BE DENIED

But of course one has to be blind not to see that outside of disease states and particular experimental circumstances muscles contract with all sorts of purposes in mind. They may contract so that I can climb a mountain, swim, knit, play the violin, or tie my shoes, each to fulfill a particular purpose. Whatever the mechanics of muscle contraction and the physics of movement may be, muscles, at least normally functioning muscles, have a motive or purpose for what they do.

Skeletal muscles, those attached to our bones, are called voluntary for just this reason. They contract at our command and for our purposes. However they behave in the laboratory, in the real world of living things, muscles work to accomplish various purposes. The mechanics and physics of contraction, as well as its underlying chemistry, offer no solace for automatism.

Nor does Occam's razor give us permission to ignore the natural world just because we can *imagine* a simpler one. However many propositions there may be, the truth must in the first instance be sought in what is observed, not by ignoring it in favor of what is imagined. That would be like asserting that a compound mathematical function is not compound at all, because not being so makes it less complicated.

Nor can we dismiss what we see by reducing it to simpler things *if in so doing we lose information*, that is, if we have to expunge some aspect of the natural phenomenon to do so. In the current case, we would have to deny purpose. The simpler explanation in this case is not merely untrue; it is not even the same thing. I can say that the engine of my automobile has fewer parts, regardless of the actual fact of the matter, because fewer parts make for a simpler machine, but I am going to have a devil of time fixing my car.

As for strong micro-reductionism, almost everything about life suggests that the whole, whether life itself or its various substituent elements in cells, tissues, organs, or systems, has attributes that transcend the parts that comprise it. For example, metabolism is more than the chemical reactions that enable it, and as said movement is more than a mechanical action. In the fullness of life's many functions, they allow or advance survival. As we shall see, the fact that living things are comprised of chemical substances and do mechanical work rather than affirming the absence of purpose in life leads to the opposite conclusion that chemistry and physics gave rise to life's purposes. Moreover, if living things evince purposes *and* are part of the cosmos, then at least certain aspects of the physical universe are purposeful.

Finally, though we cannot prove a negative, ergo the absence of purpose, if it does not exist, we should be able to show that the purposes that things are purported to serve have been misconstrued or misconceived. Appearances can certainly be deceiving. As a matter of science, we would expect that exposing such deception would lead to a deeper, more discerning and rigorous understanding of nature, say, comparing a Ptolemaic view of the heavens to a Copernican one. That is, the understanding should not merely be mistaken, but in being mistaken, it should be enlightening. As such, in denying purpose, we have to do more than merely ablate the distinction between life and rocks, we have to do more than homogenize or obfuscate what appear to be meaningful differences between them. We must per force expose a deeper truth, a deeper understanding of life's nature, as with our view of the solar system. It is not sufficient merely to rid science of a nuisance, of an inconvenience by blinkering our eyes. Even if we cannot prove a negative proposition, we have to at least be able to show that the absence of purpose in the worlds of chemistry and planets, in the world of inanimate objects does more than *permit* its absence in animate ones. The belief must be more than an analogy. It must be a necessity.

THE TELEOLOGY OF DANGER

It should be understood that introducing purpose into biology is not an argument for higher purposes. As I have described them, they are ordinary, not higher or divine purposes. They are nothing more than aspects of the physical world. This brings us to the question of final causes, *their* final cause, the final cause of living things. Remember, I said that for the teleological claim to be justified, we must be able to locate and identify the final cause of expressed purposes, most generally the final cause of survival.

If our intentions arise from within us, then at least at first glance it would seem that so would the final cause of our expressed purposes. For instance, for human artifacts it is *our* purpose to bring them into being. Are we not then the final cause of this desire, its inventor? If by lived experience or by our actions as artificers, purposes exist and arise within us, then so, it seems, must their final causes. But this is not the case. Final causes, even those of our inventions, come from the outside world. This is so even though an environmental source of a final cause may not only seem wrongly conceived but also vague and nebulous, evoking thoughts of metaphysics. This raises what at first glance is a breathtaking question: Putting godly causes aside, if the final cause of our purposes is to be found in the external world, what in that world could it possibly be? Happily, Darwin's theory provides a simple, even a mundane, all-purpose answer to this question.

As has been pointed out, the victim of a deadly assault acts with the purpose of staying alive and employs various means to that end. These purposes inhere to the quarry. The gazelle's purpose in running is to escape the cheetah, to elude the danger it presents, to escape death. Everything it does in response to the attack, whether running, jumping, going fast, changing speed, taking evasive action, increasing its heart and respiratory rate, or decreasing its gastrointestinal activity, and so forth has the common purpose of prevailing in spite of the danger, all of it with the intention of surviving.

Though survival is the gazelle's purpose, the final cause of that purpose is not found within it. Nor is it found in the attacking cheetah, notwithstanding the fact that it is the reason for the gazelle's action. It is not found in either place or for that matter in an object, but in the *danger* to survival. The cheetah does not pose a danger to the gazelle by merely being but by its action. Without the threat, without the attack, the cheetah poses no danger, and there would be no need for action on the gazelle's part, no need to run, to escape. It was the extant danger that was the final cause of the gazelle's purpose; it was the reason for its actions.

That danger, whether exigent or merely perceived, is the final cause of the actions taken by any and all organisms to survive any and all assaults. Though stress hormones and the like may flood the gazelle's bloodstream, it is the danger that elicited their increased presence that is the true or underlying cause of action. It is the final cause of the response. It is important to be clear about this, so let me reiterate: the final cause of our purposes is not found in the particular physical or chemical events that take place but in the *danger* that however constituted or represented elicits them. It is in the *meaning* of the events, not the events themselves, that we find final causes.

If I dart out of the path of a large rock as it is falling, I am of course trying to avoid being hit. But in taking evasive action, I am not merely responding to the visual stimulus of a falling object. I perceive the *danger* it presents. If a feather were falling, I would not rush out of its path. I might even attempt to grab it. Or if it started raining, I might scurry for cover, but I would not be concerned about the danger posed by the force of the falling rain. It is the particular danger posed by the falling rock with its mass, density, and velocity that is the final cause of my actions. Remarkably, in its inertness, it gives me intention. Though it has no purpose of its own, as the rock falls, it is the final cause of mine.

Nor does identifying a danger require conscious perception as in my avoidance of the rock. The danger may be made evident by something as simple as a chemical alighting on a membrane or a physical force distorting it in a single-celled organism. Even here it is not the presence of the chemical or force that is the final cause of action but rather the danger it poses. The cause of the purposeful response, the cause of avoidance, defense, or attack, is a perceived danger.

This may not seem to apply to a hungry animal seeking nourishment. After all, I eat because *I* need food, not because something external, something in the environment does. The final cause of the need to obtain, digest, absorb, and assimilate food seems internal. But even here this is not the case. We must look outside for the final cause of action.

The final cause of hunger, the need for sustenance is not found within our bodies but is instead due to the effect of the environment on us. As we live life, we expend energy and degrade and dissipate our substance to survive life's exigencies. To stay alive we must replace what we lose or use. That is why we eat. External events that threaten our existence require us to obtain food if we hope to survive. It is in the *danger* they represent that we find the final cause of the need for nourishment, the final cause of our purpose in obtaining, digesting, absorbing, and assimilating food.

And so to find final causes, we need not look to otherworldly contemplation or imposition, or to an equally unfathomable inner world. The final causes of our means of survival are found in the ordinary challenging circumstances of the world outside us. Oddly, life depends on them. I say "oddly" because this means that life depends on danger to its existence. In a world without danger, there would be no need for mechanisms to defend against, respond to, or escape its consequences. They would not exist, nor, as a result, would life.

That is to say, there would be no need for the adaptive mechanisms of life, and without them, there is no life. As explained elsewhere, it is in our adaptive mechanisms, that is, in our will to survive that life is found.[5] Counterintuitively, without the danger of extinction, life would not have come into being.[6] Though we might find biological objects in a danger-free Garden of Eden, whatever they might be, and however we might characterize them, they would not be living. Not even Adam and Eve or the tree of life would be alive without the danger of death.

We do not have to bend ourselves into intellectual pretzels to explain purposes and intentions, and their final causes. We do not need to seek them in obscure forces within us or rely on the Deity for explanation, nor do we need to deny their existence in order to be scientific. The dangers inherent in the environment impress intention upon us. The final cause of the purposeful actions of living things is *not* like devising a plan to build a chair. It is nothing more than an ultimate cause for action. Danger is the basis and cause of the action of life's purpose. This is to say that biological action is purposeful for no reason greater than the attempt to resolve the need that elicited it, to overcome the danger that is being faced.[7]

To ignore or exclude the existence of purpose in living things based on a materialist worldview may be convenient and less complicated, but it leaves us half-blind, unable to account for what nature has wrought. Purposeful actions are not figments of fertile human imaginations, the consequence of a communal delusion, or a matter of literary license. Nor can they along with their final causes be placed outside of science as Bacon said, fit for description and analysis only by the humanities.

Purpose is deeply embedded in life, in every action of every species, from flowering plants to humans. It is an essential part of the science of life. Like any other phenomenon, it is science's task to try to understand it. If an action or object has a purpose, then it must also have an ultimate or final cause. Something must provide the intention, instill or cause the purpose.

Even if some of our actions seem trivial and unconnected to survival except in the most oblique and tangential fashion, say, a meaningless utterance or hitting a ball with a bat, they are part and parcel of the system of survival. However inconsequential in a particular case, speech and hitting things can in some circumstances play a powerful role in securing our survival.

And so, if survival is the purpose, then the final cause is the danger being faced. In all of this, the source of purpose does not derive solely from life's capricious circumstances, such as the gazelle facing attack, but from the enduring demands of the rules and laws of the physical world. We now turn our attention to them and look for purpose in how physical law affects evolution.

Chapter 14

THE LAWS OF NATURE

Purpose from Physical Law

It is absurd for the Evolutionist to complain that it is unthinkable for an admittedly unthinkable God to make everything out of nothing, and then pretend that it is more thinkable that nothing should turn itself into everything.

G. K. Chesterton, *St. Thomas Aquinas*, 1933

Life did not evolve merely as the result of the action of natural selection on random mutations in light of the accidental danger-ridden circumstances of the environment. There was another influence, and it was both controlling and universal. Gravity, temperature, pressure, mass, and every other physical attribute known to us, as well as the mathematical rules that describe and underlie them, played a commanding role in life's evolution. Evolution was inescapably tethered to them; they shaped us, and there was no choice in this.

PHYSICAL LAW AND PURPOSE

This bond was manifest in two ways. The first, self-evident in modern times, is that to exist living things must conform to the purposeless conditions and constraints of physical law. The second, an anathema to many today, is also a form of compliance, but in this case compliance achieves a *purpose*. Along with their material and efficient causes, the two forms of compliance are in Aristotle's terms formal and final causes of life's attributes respectively. For instance, pond water is cooled by the wind, heated by the midday sun, and cools once again after sunset. Living things are the same in this. We passively gain and lose heat abiding by the laws of physics and its underlying mathematical rules. This is compliance according to the formal causes embodied in nature's laws and rules, and in this the animate and inanimate are no different.

The second means of compliance is unique to living things. In it the laws of physics and rules of mathematics serve as final causes of the many purposes of biological adaptations. For instance, living things do not merely passively respond to changes in temperature; we adapt. And these adaptations do not merely comply with physical law; they are *expressions* of it. We can say that the adaptation is physical law being expressed.

Without the first form of compliance, without compliance with the laws of physics and rules of mathematics, without their formal causes or foundation, there would be no basis for material or efficient causes, for static or kinetic states. That is to say, there would be no basis for the existence of matter. Without the second, there would be no adaptations. And without adaptations, there would be no life.

Adaptations evince two purposes, survival and compliance with physical law. Properties of living things that obviate or avert danger are adaptive because they provide an advantage in the struggle for survival. But they are also adaptive because of the purposes they gain as expressions of physical law. *Survival is the consequence of adaptations, while compliance is a precondition for their existence.*[1]

It is the second form of compliance that concerns us in this chapter. In what follows when I say that something about living things complies with physical law I am usually referring to adaptations. In this, I am not talking about a particular subset of adaptations that comply, but about every adaptation in every living thing, in every cell, tissue, organ, and organ system in regard to each and every thing they do, internal and external, in each and every respect, in each and every detail, from the microscopic to the macroscopic, both in their parts and wholeness, despite disparate means and disparate ends, different properties and propensities, in both first and last instances, in their earliest incarnations and their most recent. By way of illustration, here are two very general examples: heat or temperature, and movement, specifically the movement of oxygen to the cells of our tissues and organs.

THE LAWS OF HEAT TRANSFER

To mitigate heat loss at cold temperatures, animals seek shelter to get out of the wind, curl up into a ball, huddle together in groups, migrate to a more temperate clime, or for humans, turn up the heat or put on more clothing. In each of these instances the adaptation does not merely comply with the laws of physics, expressly those of heat transfer; it is an expression of them. The laws specify its character.[2]

We seek shelter to get out of the wind because the wind increases the rate of heat loss from the body; we curl up into a ball or cling together because this decreases the exposed surface area of the body, and that reduces heat loss; we move or migrate to a more temperate climate (or turn up the heat) to decrease the temperature gradient between the body and the environment, thereby reducing the rate of heat loss; or finally, humans put on clothing to insulate themselves from heat loss. Each of these adaptations not only comply with the laws of heat transfer; they are expressions of them.

Or consider blood flow to the skin. At cold temperatures cutaneous blood flow is reduced to conserve heat. Decreasing it lowers the rate of heat loss. On the other hand, when it is hot, blood flow to the skin is increased in an effort to lose heat and cool the body. Hot or cold, the adaptation is an expression of the laws of heat transfer (not to mention many other physical laws, such as those that govern fluid flow [the flow of blood]). Again, the adaptation not only complies with physical law; its makeup and properties are a manifestation of it.

Then there are our arms and legs. In addition to their obvious uses, they play an important role in heat conservation. You may have noticed that when you get cold, your hands and feet get cold first. This is because blood flow to and from them takes place by means of long parallel loops of arteries and veins, arteries that carry blood to your fingertips and toes, and veins that carry it back to the body's core, ultimately to the heart. As blood travels down the limb, heat is exchanged with blood returning from the extremities, warming it, while at the same time cooling that heading for the tips of our fingers and toes. This is an example of a physical mechanism known as countercurrent exchange. In this case, the exchange is of heat between two columns of fluid flowing in opposite directions. Such arrangements are common in animals. For example, it helps keep ducks and other aquatic birds from losing heat as they paddle in cold water, and within our bodies the kidneys use it (and another similar mechanism) not to deal with heat but to concentrate urine. In each case, the adaptation not only complies with physical law; it is an expression of it.

The most familiar example of an adaption to cold temperature is fur (hair). Though a reduction in blood flow to the skin in effect makes it a little thicker and as such a better insulator, for many mammals fur provides the major source of insulation. The effectiveness of fur as an insulator depends on two things, the length and the thickness of the coat. A long coat is better than a short one, and a coat in which the hairs are packed together tightly is more effective than one in which they are sparse. The spaces between the hairs of a long thick coat are filled with an unstirred layer of air that is warmed by the body and only exchanges slowly with the cold air of the environment. The longer and thicker the coat, the slower the rate of exchange, and consequently the more effective the insulation is. Again, the adaptation is an expression of the laws of heat transfer. It is nothing more and nothing less.

Finally, there is an adaptive response to thermal stress in which the danger is mitigated not by varying the rate of heat loss but by varying its rate of *production*. Mammals warm themselves by generating more heat, by burning more calories. This is called thermogenesis. For instance, we shiver when we are cold, that is, we contract various muscles clonically as a means of generating heat. We are not alone in this. Bees, for one example, also shiver (as well as cluster together and rotate positions from the cold edge of the hive to its warm center) to keep themselves and the queen bee warm.

In contrast, in hibernating mammals, the temperature gradient between the body and environment is reduced in cold temperatures by decreasing the metabolic rate and hence heat production, thereby reducing body temperature. The hibernating animal becomes *less* active and may even seem moribund. Paradoxically, the same sort of thing happens when ambient temperature is elevated. Animals become languid and motionless in an attempt to reduce heat production. They may lie down and spread out to increase their exposed surface area, and do so on the cool ground under the shade of a tree, a bush, or a rock, or they may actually burrow into the soil to increase heat loss. I could go on to point out sweating as a means of heat loss and subcutaneous fat for insulation. The list of adaptations that deal with fluctuations in temperature is long and varied, but in each case, whatever the adaptation, however it is incarnated, whatever particular means are employed, it not only complies with physical law; it is an expression of it.[3]

THE PUMP: THE HEART AND ITS PURPOSES

We see the same thing in the most intimate and critical aspect of our being, our heart. By now I hope you can see that the notion of a functioning but purposeless heart is silly. The heart is exactly what it is *intended* to be, a pump. But this is hardly the end of the matter. It is hardly a sufficient account of its purpose. All it tells us is what the heart is and does. We are left in the dark about why it does it. Why do we need a pump, toward what end, toward what physical or physiological purpose? We might say that it is needed to perfuse the vessels of the circulatory system with blood. But this is not of much help. We still do not know why this is necessary, what purpose it serves. It is like knowing that a chair is for sitting but having no idea of why one would wish to sit on it (for the reason, see below).

The first and most general answer to this question is prehistorical. Quite simply, the heart safeguards life. As anyone who has seen death knows, when the heart stops beating, when it stops pumping blood, death follows immediately. But still this does not enlighten us as to *why* it safeguards life. Why it is an existential necessity. Why do we need a pump? Why do we need to perfuse vessels with blood, to what end, for what *purpose*?

Nothing could be said about the heart's purpose beyond safeguarding life until the late eighteenth century when Daniel Rutherford, a student of Joseph Black, professor of medicine and chemistry at Edinburgh at the time, carried out an extraordinarily simple experiment that turned out to be one of the most profound in the annals of science. He placed a mouse in a bell jar, sealed the jar, and waited.

Before long the mouse died. It did not die of starvation or old age, but because the air in the jar was no longer able to support life. Given then-contemporary views about the relationship between combustion and life (dating back to Aristotle), Rutherford checked to see if this spent air could support a candle's flame. It could not. It could support neither life nor flame. Something about the air had changed *due to the presence of the mouse*, but what was it?

Rutherford did not answer this question, but subsequent attempts to find an answer led to a grand and enormously productive controversy. The path-breaking understandings that followed included nothing less than the discovery of modern chemistry with its elements, compounds, and reactions, the discovery of oxygen, and the realization that the mouse died because it had consumed the oxygen contained in the air. Life needed oxygen to survive and the flame needed it to burn, and there was not enough left for either. Rutherford had exposed the ancient mystery of combustion and in the process forever connected life to chemistry.

As for the heart and its purpose, British scientist Joseph Priestley, the discoverer of oxygen (or, as he incorrectly viewed it, "phlogisticated air," air that contains a substance he named phlogiston that was released from the body to prevent combustion), was the first to realize *why* the heart and blood circulation was necessary. Their deeper purpose was to supply the life-giving substance oxygen to the tissues and organs of the body (or with the advent of cell theory some seventy-five years later, to its cells). This exposed an even more profound question. Why were a pump and all the plumbing of the circulatory system needed to do this? After all, outside of bell jars the air is filled with oxygen.

DIFFUSION AND FLOW

The answer to this question had to wait for the unearthing of new laws of physics. During the nineteenth century it was discovered that nature provides two means of moving substances: diffusion and flow. Equations specifying the relevant variables and constants for both were written and their character rigorously analyzed. The heart made use of flow, not diffusion. As we all know, it produces the *flow* of blood.

But why, why was flow, not diffusion, chosen? On the face of it diffusion is far more desirable. In it elements, molecules, and substances of all sorts move as individual particles. Each object moves as the result of its own intrinsic kinetic energy (called Brownian motion). No external mechanisms, no pumps, no blood vessels, or other conveyances are needed. It is the ultimate in kinetic simplicity. Things move of their own accord. Why didn't evolution choose this far simpler mechanism, as Occam's razor teaches? Why favor the enormous complexity of the heart and circulatory system? Diffusion was not only simpler; there was nothing to evolve. There it was, ready to be used, a fundamental aspect of the physical world.

The problem was that for all its advantages, diffusion could not get the job done. It could not produce the movement required if large animals such as humans were to exist. The reason is that it is only effective in moving substances over very short, indeed *microscopic*, distances, tens of microns, thousandths of a millimeter, the width of a couple of biological cells, no wider than a strand of human hair. If the distance is greater, if it is macroscopic, millimeters, meters, not to say miles, diffusion is useless. As the distance to be traveled grows, the rate of diffusion falls off exponentially. It takes microseconds (10^{-6} seconds) for oxygen to cross the thin enclosing membrane of biological cells (10^{-5} mm), milliseconds (10^{-3} seconds) to cross cells (3×10^{-2} mm), and *hours* to diffuse across our bodies (several hundred millimeters).

Check this out with a simple experiment. Place a thin layer of a dye on the bottom of a tall glass and then fill the remainder of the glass with water, trying not to disturb the underlying layer as you add the water. If you are successful, you will see that the colored material quickly enters the clear water immediately adjacent to it, but it takes hours for it to reach the top of the glass. This is diffusion at work.

If diffusion was the only way that oxygen could penetrate the body of animals, they would by necessity be very thin, the thickness of single cells or thin layers of cells. For creatures as thick or thicker than humans, comprised of many layers of cells and interposed extracellular material, diffusion would take far too long. We could not survive the wait even over so short a distance as the thickness of a pinky finger.

Fortunately for our close relatives and us, nature provided an alternative—flow. Flow can move substances over macroscopic distances. This is the same flow that moves water and its contained substances in our rivers and streams. It does not produce movement object by object but en masse or in bulk, carrying everything from source rivulets and brooks in the mountains to the estuary as the river meets the sea. Movement in this case is produced by the difference in the elevation of the water as it falls from mountainous heights to sea level. This difference produces a (hydrodynamic) pressure gradient, ultimately due to gravity, that drives flow, just like a ball sliding down an inclined plane.

But living things cannot be in the mountains and at the seashore at the same time. Some other way was needed to produce the pressure gradient necessary to generate flow in our bodies. Nature provided it in the form of a mechanical pump. Based on the pressure the heart generates, it takes about a minute for oxygen-containing blood to circulate through all the organs and tissues of the human body, this against the many hours required for diffusion to carry substances from the body's surface to its core.

In doing this, the heart has the two purposes we have already discussed. The first is survival. The flow that the heart produces provides enough oxygen to our tissues and organs to maintain life. The second is to comply with physical law. As an adaptation, the heart is an expression of physical law. In particular, it produces sufficient flow to overcome the limitations of diffusion. Without the first purpose we would perish, without the second we would not have come into being in the first place.

The ability of the heart to move large volumes of fluid quickly over macroscopic distances allowed for another, even grander, movement, this time not of fluid but of living things themselves. The heart and blood circulation permitted the evolution of animals large enough to move over geographical distances, from mere inches to circumnavigating the earth, in an unending search for kinder, safer, and more plentiful surroundings.[4] Without them, animal life would have been limited to small creatures dependent on the will of the wind and currents of the seas for movement over macroscopic distances. For creatures like us to exist, the evolution of the heart was not a matter of chance or choice but an absolute necessity. Physical law required it.

IN THE END, DIFFUSION AND SOLUBILITY

To further highlight the centrality of physical law to the mechanisms that provide oxygen to our tissues and organs, we can look at two additional properties, not properties of the heart but of blood vessels and blood. In vertebrates, blood is of course circulated through blood vessels, arteries, veins, and capillaries in a closed circuit. As arteries carry blood away from the heart, they branch or arborize, decreasing in diameter and increasing in number. By the time they penetrate the depths of the body's various tissues and organs they have been reduced to countless tiny micron-sized capillaries. It is here at the capillaries that blood discharges its oxygen load to supply the cells of the body.

After releasing its load, the now oxygen-poor blood returns to the heart. The capillaries converge and form thin-walled veins that increase in diameter and become fewer in number. No sooner does the blood enter the heart's chambers via the central veins, or the vena cava than it is pumped out again. This time the entire volume is carried to the lungs (or gills in fish) where its oxygen content is replenished from the environment through a huge capillary bed (the role of physical law here is critical and is a whole other story [see below]).[5] Having been recharged, the newly oxygenated blood returns to the heart only to be pumped out again without surcease, this time into the arterial circulation to repeat the cycle.

For all this complexity, the flow of blood in the cardiovascular system only gets the oxygen-rich blood *close* to the cells that need it, and of course close is not good enough. To actually reach the needy cells, oxygen must leave the capillaries to travel the remaining distance to the target cells. This distance is short, indeed microscopic, and as a consequence a shift in the mechanism of movement takes place from the macroscopic world of flow to the now highly efficient microscopic world of diffusion. Diffusion becomes the chosen means of transportation.

To facilitate diffusive movement, the capillary bed evolved three critical properties. The first is that their number is very large. This provides an enormous surface area, and physical law tells us that the rate of diffusion increases in proportion to surface area. The second is that a single layer of thin or flattened cells encloses the capillaries. This minimizes the thickness of the barrier that has to be crossed to leave the bloodstream, and again in accordance with physical law this increases the rate of transport. And finally, the great profusion of capillaries ensures that the distance between any given capillary and nearby cells is very short, and this increases the effectiveness of diffusion.

Though in the microscope, the structure of many tissues looks disorganized to the untutored eye, this is not the case. It is organized with these three features in mind. What you see was not only designed to achieve effective oxygen delivery, but to do so with the quantitative precision required to meet the needs of the organism. The law of diffusion mandates each property, the number of capillaries, the thickness of their membrane, and the distance between them and the cells they serve with quantitative precision. Taken together they ensure that sufficient *quantities* of oxygen are provided to the cells being served.

The second trait concerns blood and the substance that makes it red, the protein hemoglobin. No less than the heart and capillaries, hemoglobin is vital for adequate oxygen delivery. The variable in this case is not flow or diffusion but solubility. It turns out that the solubility of oxygen in water is relatively low. If dissolved oxygen were all that were available, then everything else about the cardiovascular system, the heart included, would be for naught. There is far too little oxygen (21 percent in air) to sustain life. Even 100 percent oxygen would not be enough.

To overcome this limitation, a protein molecule with a special affinity for oxygen—hemoglobin—evolved. It binds oxygen and enables blood to carry much more than would otherwise be possible. Along with the heart and the capillary network, hemoglobin was necessary for the evolution of even the smallest vertebrates. Without it or some similar molecule they simply would not exist; they could not have evolved. Here again, evolution was not just a matter of chance but necessity. Though, like all proteins, the structure of hemoglobin's antecedents changed as the result of random DNA mutations, what evolved had the specific purpose of making it possible for blood to carry enough oxygen to fulfill the needs of large animals for this substance. Once again, physical law imposed a purpose, in this case not on an organ or tissue but on a molecule. Hemoglobin's structure is an adaptive expression of physical law.

But this still was not enough. Hemoglobin had to be present in the necessary amount. There had to be a particular value for the product of the number of hemoglobin molecules and the avidity with which they bind oxygen. Evolution had to yield a mathematical product of these two variables whose value was sufficient to supply enough oxygen to meet the body's needs. Both the amount of hemoglobin that the body produces and how avidly it binds oxygen are at one and the same time adaptive properties of the molecule and expressions of physical law.

And as with temperature regulation, we have hardly exhausted the adaptations needed to supply oxygen to our cells. For instance, simply carrying oxygen to the site where it is needed, even with all the features just discussed, doesn't get the job done. It must be *unloaded*. It must be released from hemoglobin and in the right amounts to maintain kinetic and thermodynamic balance between what is provided and what is used. In accordance with physical law, mechanisms evolved that do just this (I will spare you the fascinating [to me] but complex details). As it happens, every feature of the cardiovascular system, property by property, mechanism by mechanism is both an adaptation and an expression of physical law.

Even with all this, this only tells half the story. The other half is how oxygen is extracted from air (or water). That is the job of the lungs (or gills) and the processes known as respiration. Once again, each adaptive feature, each mechanism, each property is an expression of physical law. I will leave it to you to consult the relevant literature to see how this works if you are interested.[6] And of course oxygen availability, however critical, is not the whole of life. We can apply the same analysis with the same kind of results to the acquisition of food, to defending oneself from becoming food for others, to the removal of harmful waste (excretion), and so forth. Each adaptation is ultimately an expression of physical and chemical law. It does not merely apply it; it is physical law actually being expressed.

A CHAIR FROM THE COSMOS

The case for purpose made here and in the previous chapters is of course a response to the argument *against* teleology. It declares that biological adaptations have inherent purposes that arise from final causes found in the material world, found, as explained, in environmental danger and physical law.

One aspect of life appears to stand in stark contrast to this narrative—human inventions and constructions. In this case, the inventor appears to be the final cause of the invention's purpose. Consequently, the final cause is internal, not external, within us, not in the world outside. As already touched upon, this is a misconception. Yes, the inventions are ours and we impose our purpose on them, but their final causes are another matter. They, it turns out, like the other adaptive purposes of life, arise from the external world.

Though we are builders and designers nonpareil, and no other species can claim more than a barely visible shadow of human imagination and resourcefulness, as naturalists have pointed out time and again, inventions and constructions are not restricted to humans. Birds of course build their nests, bowerbirds their bowers, beavers their dams, bees their honeycombs, spiders their webs, and ants their hills. In contemplating *their* cause, whether it is for protection, food storage, sex, or capturing other creatures, we do not evoke some unique "birdish" or "antish" internal final cause but rather the same two quotidian external causes we have been talking about, danger to life and the laws of physics. Human inventions are no different.

For example, chairs are not merely something that someone thought up—"it would nice to have chairs." They provide gravitational support. Sitting on them reduces gravitational stress on our vertebral column and leg bones and provides rest for our postural muscles in ways that have various practical advantages over lying or sitting on the ground. And of course, gravity acts externally.

If it were not for gravity, we probably would not have chairs. There would be no need for them. We would do the things we normally do sitting down standing up. Why not eat and rest standing? Indeed, in the absence of gravity, to sit we would have to tie ourselves down to chairs, as astronauts do.

And so though the invention and construction of chairs reflects human desires and intentions, their ultimate cause is not some inherent human craving for chairs, some special internal human final cause of "chairness," but the force of gravity.[7] However ingenious, however well planned by human inventors, chairs are the result of causes that originate in the physical world. They impose their will on us and we resourcefully comply.

We can supply similar attributions for all sorts of human artifacts. For example, we manufacture clothing and housing to provide protection from the heat, cold, wind, and rain and invariably seek to do so in ways that comply with nature's laws, beginning, but not ending, with those of heat transfer (the "Three Little Pigs" discovered the necessity of compliance the hard way). Or we invent weapons, such as knives and guns, in strict accordance with the laws of physics. We do not make knives out of feathers or use feather bullets in our guns. If we did they would be of little use in attacking or defending against attack.

Together survival and compliance with physical law are the underlying purposes of all our creations. Whatever dangers they are meant to elude, preclude, or forestall, and whatever physical characteristics they express, and however shrouded and obscure they may be, all our adaptations can be attributed to these external causes. This is even so for *art*, written, performing, musical, or visual, profound, dreadful, or blatantly idiotic, from Shakespeare to the silliest of sit-coms. However remote or distant, a connection to environmental danger and physical law can invariably be made.

This makes humans one with the rest of life, with our closest relatives and our most distant cousins alike. We are not the inexplicable exception, a source of hidden final causes, God in disguise. Though there is still the remarkable and unique ingenuity of our species to account for, it appears that the final causes of human artifacts, of our inventions, do not come from within us but, however veiled and murky, from the world we inhabit.[8]

THE FALLACY OF LIFE WITHOUT PURPOSE

This brings our general discussion of purpose to an end. We have learned that the common belief that life's evolution is merely the consequence of undirected, *purposeless* processes; of chance heritable changes, or mutations; and haphazard environmental circumstances is incorrect. Life is fully endowed with purpose, whether it concerns survival or compliance with physical law.

The final causes of these purposes are twofold—the existential *dangers* we face and the physical *constraints* that nature's laws impose on us. The former instill the purpose of survival, while the latter demands compliance in order to express life's adaptive features. Reproduction aside, they form the basis of biological evolution. Our adaptations emerged in an attempt to both survive and comply.

The laws of physics and the rules of mathematics not only invested living things with purpose but are responsible, decisively and formatively, for making the remarkable anatomical, biochemical, biophysical, and physiological properties of life, so beautifully tailored to their appointed tasks, what they are. Whether to retain or purge heat or provide oxygen to our cells and tissues, every adaptive property of life is an expression of physical law, an essential and inescapable part of the search for adaptive meaning and merit.

There is nothing remotely otherworldly in any of this. To the contrary, whether due to environmental danger or the laws of physics and mathematics, final causes arise from the ordinary properties of the material world. They come from the same disinterested forces of nature on which natural selection depends. Biological purposes do not exist in opposition to natural selection's agency but rather in harmony with it. Purpose, it turns out, is not a problem for Darwin's theory. It provides substantiation for it. Materialism and teleology are boon companions.

After all is said and done, we cannot conclude, as my professors did all those years ago, that the tissues and organs of our body evolved to perform their functions in a purposeless fashion. This said, you might have noticed an enormous omission from the discussion to this point. What has been missing is nothing less than the principal subject of this book, reproduction and *its purposes*. As explained, this was my intent. The object was to first consider purpose in regard to life's somatic features to lay the groundwork for a discussion of reproduction and its purposes, to which we can now turn our attention. Can we also identify *its* purposes and final causes in the material world?

SECTION V

PURPOSE AND REPRODUCTION

Chapter 15

THE PURPOSES OF REPRODUCTION

Why Reproduction?

It ought to be remembered that there is nothing more difficult to take in hand, more perilous to conduct, or more uncertain in its success, than to take the lead in the introduction of a new order of things. Because the innovator has for enemies all those who have done well under the old conditions, and lukewarm defenders in those who may do well under the new. This coolness arises partly from fear of the opponents, who have the laws on their side, and partly from the incredulity of men, who do not readily believe in new things until they have had a long experience of them.

Niccolò Machiavelli, *The Prince*, 1505

Whatever misgivings and uncertainties one might still harbor about the idea of purpose for the somatic features of life, reproduction and its features are thought not only to have a purpose, but that purpose is considered self-evident, common knowledge—of course, I am referring to making children, creating progeny. And yet if we believe in a non-teleological world, in a world without purposes and final causes, this cannot be correct. Reproduction cannot have a purpose. It can only do what it does. Sex takes place, leads to reproduction, and reproduction begets progeny, all in the absence of purpose.

Envision if you will an Alice in Wonderland assembly line built without any intention of doing so, and that having been built unintentionally manufactures a purposeless machine. Both the mechanism that produces the machine and the machine it produces exist without reason, intention, or desire. There is no Aristotelian purpose to any of it. It is all the transactional consequence of disinterested and uninformed causes.

This is reproduction in a world without purpose. Though its means and product would be the result of evolution, not our imagination, just as with our Alice in Wonderland concoction, in such a world the coming into being of new living things would be without motive or purpose. As peculiar as it may seem, this is apparently how Aristotle viewed reproduction. Remember that he said that the father is the efficient, not the final, cause of the child, and efficient causes lack purpose. They just make things happen. For example, in causing an inanimate body to move, a mechanical force is the efficient cause of its movement, but it intends nothing in this. It is aimless. Only final causes generate purpose.

RANDOM CAUSATION

A purposeless mechanism of reproduction emerging from biological evolution would have been an incalculable oddity. It would have required useful though randomly achieved mutations to accumulate and favorable though chance environmental circumstances to govern, despite the presence, indeed the predominance, of countervailing useless or disorganizing mutations and unfavorable environmental circumstances. Traits that favor the emergence of the mechanisms of reproduction would have in some causeless way been gathered together while those contravening its materialization would have been barred.

This violates the second law of thermodynamics and its call for disorder and randomness. Life's varied reproductive mechanisms, from flowering plants to mammals, would be the product of some vast incongruous chance happening in which DNA is faithfully replicated and the organism dependably copied. If Aristotle truly imagined reproduction this way, he was imagining not merely the unlikely but the unattainable. Efficient causes are not sufficient to give rise to reproduction. For its mechanisms to have come into being and do what they do they must embody a purpose, and in embodying a purpose they must have a final cause. There is no way around it.

THE PURPOSES OF REPRODUCTION

Yet shockingly and unmistakably that purpose is *not* producing offspring. That can no more be the purpose of reproduction than eating is the purpose of eating, or breathing is the purpose of breathing. Reproduction being its own purpose would be tautological, redundant.

Naturally, the purpose of eating and breathing is survival. But what can we say about reproduction's purpose? If it is not producing offspring, then what is it? To what end is new life created? Or put another way, if reproduction were a means of producing living things merely so that they could live, and that to live is to be endangered and ultimately to die, why go to all the trouble—and trouble it is—to produce more? Why have children, why does nature create progeny, for what purpose, toward what end?

There is a purpose; it is just that that purpose is not the progeny themselves. The purpose of reproduction is the continued existence of life and its reproducing groups, its family lines and species, not the individuals it creates. This is a spectacular, even a grand, purpose. As individuals, we are of course doomed, our existence is fleeting and reproduction does nothing to change that. But at the same time, by its very nature, reproduction extends life beyond our existence. It ensures our patrimony. Just as mechanisms of survival are attempts to secure the present for individuals, the mechanisms of reproduction are attempts to guarantee the future of life's diverse reproducing groups and their evolution. Reproduction is not about what is but about what is to be. It is about ensuring the future for living things, about ensuring the future of life.

NATURAL SELECTION AND REPRODUCTION

As with life's somatic features, for the reproductive features of life to have evolved, some sort of selection, some sort of value judgment between different embodiments had to have taken place. Without it, there could be no evolution. However, as we have discovered, the processes of reproduction, those that precede fertilization, such as sexual attraction, sex, meiosis, spermatogenesis, as well as those that follow, most importantly fetal development, cannot be attributed to the concatenation of purpose, final cause, and agency—survival, danger, and natural selection that is responsible the evolution of the somatic features of life. Some other selective means must have been involved.

Since reproduction encompasses an enormous variety of mechanisms, the fact that natural selection is not the agency of their genesis or the source of their purpose cannot be considered an inconsequential exception to Darwin's theory. In fact, it appears to wipe out the theory's universal character. Faced with this enormous threat, it is not surprising that over the years the theory's adherents and devotees have held that whatever arguments draw us to exclude natural selection as the agency of the evolution of the reproductive mechanisms of life, however persuasive and informed the reasoning, must somehow, in some way be wrong.

The evolution of these mechanisms *must of necessity* be the consequence of natural selection, just as with life's somatic features. Indeed, as said, this has been the common wisdom since Darwin's time. In this view the theory leaves no room for some other process. But this puts us in a terrible pickle. We have to somehow fit or, if you will, squeeze the evolution of the reproductive features of life into the theory of natural selection, not, as one might expect in science, the other way round, with the theory accommodating the facts.

FITTING FACTS TO THEORY

As already touched upon, there have been three approaches to reconciliation with Darwinian theory. The first argues that, despite what I have said, the processes of reproduction not only have survival value for the parents; this is their raison d'être. Their value is not found in the creation of new life but in providing an adaptive advantage to the parents, an advantage that is not expressed during the events of reproduction, but after they are over, *after birth*. The idea is that the presence of offspring alters the environment to their parents' advantage. For instance, as said, progeny can serve in combat in their parents' stead, they can obtain food for them, or they may simply be alternative fodder for destruction by others.

Though there are many examples of parental advantage, when life's entire scope is taken into account, they are little more than minor exceptions to a general rule that progeny do *not* serve the needs of their parents. As already explained, reproduction can be a dangerous business for the parents. Beyond the perils of sex and childbirth, offspring often *compete* with their parents for food, space, and safety.

But even more important, in many, perhaps most species, progeny neither help nor hinder their parents. Their birth is without effect on them. They may not even be in the same environmental niche. For example, the female may release or deposit her fertilized eggs and abandon them forthwith, as with most egg-laying fish. And for his part, the male may desert his progeny right after copulation. As a generalization the "parental" approach to accommodation with Darwin's theory seems hopelessly flawed, if not completely frivolous. Such things do happen, sometimes offspring provide important advantages to the parents, but unmistakably, reproduction's purpose has to do with producing a new generation, not the survival of the old one.

The second line of attack has probably been the most widely followed. In this view, the adaptations of reproduction do not benefit *individuals*, parents or offspring, but rather the *group* to which they belong, their family line or species. In this, it is the *group's survival*, and only dependently that of its individual members, that is at stake. Critically, it is not the other way around. The group benefits not because its members benefit, but it benefits as an entity unto itself. It is on its fate, the fate of the group, that the choices of natural selection bear.

But such a process of natural selection is just not possible. Natural selection can only work its wonders on *material* things, and groups are not material things but, as explained, abstractions, mathematical sets. They are certainly affected by natural selection, but only as a consequence of what happens to individual members of the cohort. Unless natural selection can in some fashion act on groups as objects in their own right, as if they were things, this approach to accommodation with Darwin's theory also fails.

This leaves the third and final approach. Accommodation is sought by holding that reproduction's environment is internal, not external. For example, the mammalian mother is the environment for her fetus, as the egg is for the fetus it encases. Some scientists believe that the essential environment is no more than the fetuses' own DNA. After all, it houses the genome that contains all the information needed for life's materialization.

But how would this work, the selection of what against what? Of course, for selection, and hence evolution, there must be things to select among. But such choices do not exist in the world of reproduction. Features of the individual fetus do not come in pairs that are in some sort of competition with each other. To the contrary, even for such things as bilateral or radial symmetry or the duality of DNA, we are talking about copies of the same thing there to complement, not compete with, each other. Anyway, competition or not, choices within an individual fetus can only affect is *its* viability. There is no choice that takes place between individuals akin to those of natural selection.

When all is said and done, all three approaches fail and an accommodation to Darwin's theory and its natural selection just cannot be made. I have revisited these tactics once again to acknowledge that whatever our beliefs, however strongly held, we have reached the end of the line. There seems to be little choice but to put Darwinian theory and its natural selection aside when it comes to the evolution of the reproductive features of life. It is just not responsible for their evolution. That job lies elsewhere, with some other agency.

THE MEANS OF SELECTION

Self-evidently evolution would not have taken place without reproduction. Life would end with the first generation. In addition, sexual reproduction, with its mixing and sorting of the parents' genes (along with mutations) produces progeny with the varied and distinctive characteristics necessary for natural selection. As a consequence of both, reproduction allows natural selection to act as the agency of evolution for life's *somatic* features. The curious thing is that it does not play this role for itself, for the evolution of its own features. It does not permit natural selection to act as the agency of selection for its features. But how then did they evolve? How did reproduction's mechanisms and their purposes emerge?

We have identified two purposes for the mechanisms and processes of life's *somatic* embodiment: to survive and to comply: to survive the dangers of existence and to comply with physical law in the expression of its adaptations. Their final causes are ordinary properties of the world in which we live. For survival, they are any and all causes that present a danger to life, while for compliance they are the laws of physics, chemistry, and mathematics. The question is whether a similar duality of cause and effect exists for life's reproductive features. If so, what is it, and if not, what takes its place? By what means did the mechanisms of reproduction evolve? In the next chapter we try to answer this question.

Chapter 16

THE MEANS

The Agency of Reproductive Selection

It's hard work to find a mate!

Elijah Zinn Boyd-Rothman, five years old,
on observing butterflies

Evolutionary biologist Richard Dawkins explains the miracle of natural selection in his 1996 book *The Blind Watchmaker* in terms of a popular incarnation of older concepts about probability and randomness known as the infinite monkey theorem. The theorem says that given enough time an eternal monkey or group of monkeys aimlessly banging on a typewriter or group of typewriters (or should I say computer or group of computers?) would produce Shakespeare's collected works. Given more time they would produce all the masterpieces of literature, and for that matter with sufficient time they would produce everything lexical, present and, amazingly, even that yet to be written.

Sooner or later Shakespeare would emerge from the industrious monkeys as they indiscriminately strike the keys. If we ignore restricting factors, such as their fatigue, the time needed to press the keys, and the wishes of the monkeys, if we know how many letters a work contains, the monkeys could not only produce it; we can determine how long it would take them. But don't start gathering monkeys for the job. It would take more time than we have available.[1]

In recent years, this idea has been employed to explain everything from thermodynamics (e.g., gas molecules), to quantum mechanics, to the uniqueness of human cognition, and of course, in Dawkins's case, to biological evolution. Dawkins describes how a mechanical selection like that envisioned in natural selection affects the time required to produce a short sentence by random generation. Without selection, the time needed to create even a small sentence using the twenty-six-letter English alphabet is forbidding. But amazingly, simply by inserting a selective process, the sought-after sentence can be produced in relatively short order.

It works something like this: An array of random sequences of letters of the same length as the sentence is generated. From among these sequences those that display the correct placement of one or more letters are chosen for further iteration. A new array of random sequences is then produced for the remaining letters, and once again those closest to the sought-after sentence, now containing several letters in the correct location, are selected, and once again the unstipulated letters, now fewer in number, are subjected to the generation of random sequences. This process is repeated over and over, generation after generation, selection upon selection, and before you know it, voilà! The sought-after sentence. With the tool of selection in hand, a task that otherwise would take almost forever is accomplished relatively quickly. And all we have done to achieve this miracle is tether random generation to selection, and this is of course exactly what natural selection does. It connects random mutations and genetic rearrangements to a selection for survival.

Still, sentences are not life and there are important differences between this exercise and life's evolution. For instance, the linguistic selection had the particular goal of generating a sentence that we had chosen, while natural selection has no specified goal other than survival and the limits to existence imposed by physical laws and mathematics. Even so, the theorem can provide some estimate of how long it would take for a living thing to emerge by chance compared to the time actually required.

The number of words in biology's glossary is thought to be the number of genes, understood today to be segments of the DNA molecule that code for proteins. Assuming this is correct, we can estimate how long it would take for life to materialize by chance. For example, a bacterium that contains 200 genes would be generated spontaneously at a frequency of $(1/200)^{200}$ or 1/200 times 1/200, two hundred times. This is of course an exceedingly small number and tells us that the likelihood of even a simple bacterium forming by chance is almost nonexistent. But however remote, compared to the time required to produce a human with 20,000–30,000 genes, spontaneously forming bacteria would be downright commonplace. For us, the frequency would be an unimaginable $(1/20{,}000)^{20{,}000}$. What it actually took was a measly 4 billion years.

A REPRODUCTIVE SELECTION

This shows the power of natural selection. It takes the well-neigh impossible and not only makes it possible; it makes it likely, even inescapable and does so in relatively short order. Through the intervention of choice, order is imposed, the vector is entailed, and the arrow of direction impressed on what would otherwise be random events. As explained, without an agency of selection, the mechanisms of life could only have arisen as the result of some strange providential alignment of haphazard and inordinately unlikely events. Such an occurrence would certainly rid us of the need for a purpose and a final cause, but then biological evolution would be almost as likely as all the gas molecules in a room settling in one corner. Though it could happen, you shouldn't stay up all night worrying about a lack of oxygen. The chance during your lifetime, the life of the universe, or for that matter the life of many universes borders on the nonexistent.

This tells us that in rejecting natural selection as the agency of reproductive evolution, we have not relieved ourselves of the need to identify some sort of selective process to account for it. As suggested, selection (among disparate incarnations) remains a necessity for any kind of evolution. *In lieu* of natural selection some other selective process, some other means of connecting cause to purpose is needed. In the case at hand, there must be a *reproductive* selection, as distinguished from a *natural* selection.

Though both involve selection, the choices of the two are otherwise as dissimilar as can be. Most importantly, their objectives are totally different. As explained, in the face of the brief life of individual organisms, the purpose of reproduction, and accordingly that of the evolution of its mechanisms, is to ensure the continued existence of life's various reproducing groups, and through them the continuance of life more broadly. To achieve this end, reproductive selection is not concerned with survival value, as is natural selection, but rather with procreative worth. Features that are advantageous in producing offspring are retained, while those that are less so or that are detrimental fall by the wayside.

This may seem like nothing more than a semantic distinction. After all, both processes of selection are ultimately about survival, about the fate of living things. Survival, continuance, what is the difference? However, the difference between the two is not just a matter of words. It is substantive, decisive, indeed, defining.[2]

Selection for the somatic features of life concerns "being" in that the survival of existing beings is paramount, whereas selection in relation to reproduction is about "coming to be," about future life. We can say further that natural selection is about life's uncertain fate in the face of danger, while reproductive selection is based on secure knowledge of that fate—calamity and death. An appreciation of life's ultimate fate is the reason for reproduction. For life's somatic aspects death is a *probabilistic eventuality*. For reproduction it is *foreseen knowledge*.[3]

WHAT IS SELECTED FOR?

The two forms of selection differ in another way. For its part natural selection is all-embracing. It acts on properties of living things that are as varied as the experience of life itself. Whereas reproductive selection is not only focused exclusively on reproduction, it is limited to eight stipulated features:

- The enticement that brings about sex.
- Sex.
- Fertilization.
- Protection of the fetus.
- Fetal development, the actual generation of new beings.
- The birth of a new creature.
- The number of offspring generated with each reproductive iteration.
- The number of conceiving events in a lifetime.

In most species, the role of the male, as opposed to its seed, excludes the fourth through the sixth property—protection of the fetus, fetal development, and birth, whereas the female is commonly involved in all of them. Progeny for their part bear primary responsibility for their own (fetal) development. Whatever the particular reproductive situation, the triumvirate of male and female parents and nascent offspring work toward a common goal—producing the newborn creature.

These eight elements form the general basis of reproductive selection. I say the *general* basis because they apply to all species and all reproductive processes regardless of their physical embodiment and environmental circumstances. For instance, even though each species has sex in its own way, in a way that fits its particular physical incarnation, its specific shapes, indentations, protuberances, and emanations (for instance, odors and sounds), as well as its distinctive environmental circumstances, the same eight elements apply.

This said, even though the bases of selection are the same across species, the embodiments that allow sex and result in reproduction not only differ; they are unique to each species. The reproductive features of life not only differentiate species; they are exclusionary. That is to say, in most circumstances reproduction can only take place between members of the same species. In this light, a catalog of the various means of reproduction is roughly coextensive with an inventory of species, and perhaps more than anything else about life illustrates its amazing diversity.

REPRODUCTIVE ADAPTATIONS

Reproductive adaptations are traits that improve the effectiveness of the eight features of reproductive selection. For the first six they are qualitative (though, as we shall see, they have critical quantitative effects). They are about ensuring the successful completion of reproduction per se. The last two are quantitative and are about producing enough offspring to guarantee survival of the group.

For the qualitative adaptations, selection acts to produce a reproductive system that:

- carries its task to completion,
- does so reliably time after time,
- creates an organism that is faithful to its kind; that is an approximate copy of its progenitor, and
- limits procreative sex to members of the opposite gender within a particular group of organisms.

In other words, as a qualitative matter reproductive selection seeks a system that is reliable, faithful, and restricted.

In quantitative terms—the number of offspring produced with each iteration and the number of conceiving events in a lifetime—selection has only one goal and one adaptive property, and they are in a sense one in the same. The goal is to produce enough progeny to ensure the group's continuance. And broadly speaking the adaptive property is that the number of offspring that reach sexual maturity and reproduce be equal to or greater than the number in the prior generation. The latter is required to achieve the former.

A group may produce only a few offspring as in large mammals, or many, as in all sorts of invertebrates and plants, but however large or small their number, each generation must be large enough to replace its predecessor. If it does not, then the group's presence in the population overall will decline, and if this continues generation after generation the group will ultimately become extinct. An effective reproductive system must do more than produce progeny. It must produce enough for generational replacement.

BEFORE

Natural selection and reproductive selection also differ in regard to when and where they take place. I am not referring to the time of year or the environment but to the occasion. For natural selection, it is a time and place of stress or trauma. For reproductive selection, the corresponding time and place would of course be during reproduction. If selection preceded reproduction, it would be about something that has yet to take place, and if it happened afterward, it would be about something that had already taken place.

Yet however self-evident this may seem, the opposite is the case. Reproductive selection does *not* take place during reproduction but, however incongruously, before and after. We can understand why by asking what selection would look like if it took place during reproduction. What would be selected for or against? As already noted, the reproductive features of life are not in competition with each other; they are complementary, united to achieve a common goal.

As a consequence, there is no choice to be made among its elements, say, between sex and fetal development. And certainly there is no rivalry between male and female sexual organs. The production of eggs and sperm are not in competition with each other. Now that would be a fine mess. The simple fact is that there is no opportunity for selection during reproduction because its mechanisms and processes take place within *individual* organisms, some in the male, others in the female, and still others in the fetus. For selection to occur, the entities must be in some sort of competition with each other. As it happens, for reproduction, competition occurs, and only occurs, between breeding organisms, and that can only happen before or after reproduction. *They compete with each other in terms of their procreative success both prospectively and retrospectively.*

Darwin's theory of sexual selection is about prospective selection, about choosing a partner. This is usually imagined to be the result of an assessment of the *fertility* of a potential mate based on an appraisal of certain external features—shapes, coloration, ornamentation, and emanations. Darwin pointed to the size and elaboration of antlers in male deer and the beauty of feathers in male birds.

But trying to gauge fertility on the basis of external features is an uncertain business. For instance, in humans we can imagine that features that express power and grace are markers for male and female fertility, respectively. However, experience tells us that a striking appearance, however alluring, is no guarantee of fertility. The problem is that however compelling and conspicuous the appeal may be, the choice is made *in the absence* of express and detailed anatomical, gross and microscopic, physiological and chemical knowledge of the reproductive system of the prospective partner, that is, without direct knowledge of his or her actual fertility. However evocative and redolent the choice, it can never be more than a guess.

Beyond this, in most species all individuals capable of having sex have it and procreate whatever their allure or lack thereof. In fact, in many species the choice of a partner is not a matter of preference at all but chance, as in cross-pollinating plants. And so, the features of sexual attraction, though highly evolved and sometimes of remarkable power and incomparable beauty, are weak and uncertain as measures of fertility. They are at best good guesses, at worst, lousy guesses. Given this, it seems far more likely that sexual attraction is there to help ensure that reproduction occurs *regardless of the fertility of the union*.

AFTER

If sexual attraction as a measure of fertility is little more than speculation, what happens in the aftermath of reproduction is anything but. It is clear and concrete. It not only provides direct evidence of fertility; it supplies its ultimate quantitative measure—the number of offspring the mating pair produces. Given the uncertainty surrounding the meaning of desire as a prelude to sex and the certainty of the meaning of the number of offspring produced in its wake, it should come as no surprise that Darwin's theory of sexual selection has either been ignored or been a matter of dispute since its proposal almost 150 years ago, while the number of progeny that are produced has served as the foundation for the study of the genetics and evolution of populations of organisms for over three-quarters of a century. Whether or not a meaningful measure of fertility takes place before reproduction, the principal basis for reproductive selection is found in its aftermath.[4]

This leads to a rather lengthy but instructive list of deductions:

- Because selection occurs after the progeny are created, reproductive systems can only be compared to each other as a whole, that is, without direct reference to the details of their incarnation.
- As such, the measure of reproductive selection is the number of offspring.
- Accordingly, it is based on the fertility of mating pairs.

Further,

- The number of offspring is a variable product of the mechanisms of reproduction.
- As a result, the number of progeny is not only a *measure* of selection; it is its *basis*.
- Consequently, this means that if the six *qualitative* criteria of reproductive selection of the eight listed above do not undergo selection prospectively, as with traits of sexual attraction, they can only be selected for or against on the basis of their effect on the number of progeny, that is, in *quantitative* terms, such as the percent of successful birthings.

Additionally,

- Unlike natural selection, reproductive selection is an affirmation; more is always better, less is never more.
- To the degree that a particular parent produces many children, it contributes to the group's advantage, while a relatively unproductive parent reduces it, but neither parents nor progeny themselves are advantaged or disadvantaged as a result.
- Unlike prospective selection, retrospective selection is based not on differences in the fertility of individuals but on that of the group as a whole.
- When the group fails, so do all its members, however fertile or infertile.
- Also, unlike natural selection, it is the success or failure of the group, not its individual members, that is assessed.
- Remarkably, this gives material meaning to the mathematical abstraction of "groups."

In all this,

- The mutations and genetic rearrangements that alter the character of the reproductive system are subject to selection not in their own right but only as part of the reproductive system as a whole.

- As a result, genotypes and phenotypes diverged over evolutionary time as dissimilar mutations, and genetic rearrangements accumulated within each reproducing group.
- This led to a *reproductive* isolation even for organisms of the same species situated in the same environment.
- The result was speciation.
- And finally, amid this divergence, the major features of *cellular* reproduction remained essentially common to what were otherwise exceedingly diverse mechanisms, suggesting that changes in the processes of cellular reproduction are often severely or acutely deleterious.

This is all very different from natural selection. In natural selection the failure or inadequacy of particular traits or specific incarnations of traits serve as the basis for the demise or disadvantage of the whole organism. If an animal is vulnerable to heat, we know to look to the mechanisms of heat dissipation and production for the cause of its death. If it moves too slowly, we know to look to the mechanisms of locomotion for its powerlessness to escape a predator. For these and innumerable other features, natural selection is about the relative success of particular adaptations in helping organisms overcome the environmental challenges they face.

Entire systems, say, the digestive or respiratory system, are *not* subject to natural selection. Selection concerns specific traits and their effect on the system or body overall. In reproductive selection, things are topsy-turvy; they are the other way around. It is the system in its entirety that is subject to selection. It, *not its substituent elements*, is favored or found wanting. And so, the adaptations that result from reproductive and natural selection are not merely different in that one is concerned with reproductive features and the other with somatic features; they differ in this even more fundamental way as well.

DIFFERENT OCCASIONS

To this point I have ignored a critical part of the story. Reproductive selection based on relative fertility does *not* in itself establish a group's ultimate success or failure, its ability to replace itself or expand its dominion. It is the number of offspring that *survive* to sexual maturity and reproduce that determines this. Needless to say, if they all perished before they could reproduce, it would not matter how many offspring there were; the reproducing group would be gone forthwith.

To the extent that their number dwindles during maturation, this is of course *not* the consequence of reproductive selection but of the life lived; that is, it is the consequence of natural selection. This explains why a group that produces only a few offspring can be successful, while one that produces many may become extinct. Simply, the former is more successful in overcoming life's challenges *after* birth or faces fewer of them.

If the rate of destruction is low relative to the rate of production, if the difference is large enough so that the number that persists is sufficient to replace the last generation or increase its size, then we can say that the reproductive system is successful, *and this success is the result of both reproductive and natural selection*. Fertility determines the favored reproductive system *by itself* only if the competing groups face the same environmental challenges and are equally equipped to handle them.

What reproduction and reproductive selection do is determine the size of the population *offered* to natural selection for its ministrations. It creates the descendent group and determines its initial size. Natural selection then determines how many are culled on the road to sexual maturity and hence how many will be available to reproduce.

And so we find our dear friend natural selection front and center once again. In the end, it cannot be denied. It has a central role in reproductive evolution after all. If the environment is harsh, if natural selection is severe, few organisms will reach sexual maturity. As a result, there will have been a selection not only against these organisms but also against the reproductive system they exemplify.

Through its ablutions, reproductive selection improves the ability of existing organisms to produce new ones, whereas natural selection improves their ability to withstand environmental challenge. In this, reproductive selection has nothing to do with the survival of existing beings, while natural selection for its part has nothing to do with producing new organisms. Hence, life's evolution is the result of two separate processes—the negative or destructive force of natural selection *and* the positive or creative force of reproduction. In the former nature displays its disregard for life, winnowing the less adept in the face of life's harsh emollients, whereas in the latter, in reproduction, by producing a brand-new generation of organisms, it displays an abiding interest in life and its continuance. Together they determine the group's size in the larger population and ultimately its fate.

COEVOLUTION AND NATURAL SELECTION

As reproductive evolution progressed, the reproductive features of life became more efficient and reliable, more likely to produce viable offspring. At the same time as this occurred, diverse species with disparate anatomical and functional *somatic* traits were created to allow for survival under a wide range of environmental conditions. These parallel events required changes in the means and modes of reproduction to ensure their compatibility with life's evolving somatic incarnation.

Reproductive traits did not merely evolve in their own right, on their own terms, but also to be at home, to function successfully within the species' somatic embodiment and to be suited to its way of life and habitat. This applies to cells as well as to complex multicellular organisms. The mechanisms of cellular reproduction, DNA replication, cell division, cell fusion, and so forth, evolved to be effective without impairing the cell's nonreproductive activities, its metabolic and physical actions. Beyond the cellular level, this obligation explains why reproductive processes are often so strikingly different between species—say, between mammals and green plants. Whether subtle or spectacular, their divergence represents adaptations to the somatic properties and environmental circumstances of the species.

Along with isolation, these differences determined with whom the sex act can be consummated, fertilization achieved, and progeny produced. Indeed, this is the limiting factor in how we define species. For all the phenotypic differences between giraffes and elephants, between clams and lobsters, ultimately it is the anatomical, physiological, and genetic consanguinity of the sexual and reproductive apparatus of different organisms that determines both their sexual partners and species.

But if the somatic properties of life are a guide in *reproductive* evolution, serving as its setting and environment, the opposite is also true—the somatic properties of life also evolved dependently, constrained by the character of reproduction. An organism that is very well adapted to its environment, a paragon of survival, that cannot reproduce effectively, would of course be lost to evolution.

And so, the somatic and reproductive aspects of life evolved both independently and in concert, each fitting as best it could into the requirements of the other. While reproduction had to accommodate to somatic evolutionary change, so the rest of the body had to oblige reproduction. An extreme imbalance between the two would lead to the extinction of the species due either to its inability to survive or its inability to reproduce. Harmony was vital. Without it, the number of progeny would be reduced, and, if extreme enough, reproduction might cease altogether. A difficulty with reproduction would lead to the production of fewer offspring, whereas a somatic failing would mean that fewer organisms survived to sexual maturity to produce offspring. Either way, sparse results would reflect the inadequacy of the group as a reproducing unit. The cure for a reproductive difficulty would involve reproductive selection, whereas the cure for a somatic difficulty would involve natural selection.

We can think of this as a kind of "coevolution." As the term is commonly used, it refers to the paired evolution of different species. In his 1877 book *The Various Contrivances by Which Orchids Are Fertilised by Insects* Darwin provided many examples, most prominently the coevolution of insects and flowers, as well as that of certain birds, hummingbirds and the like with bird-pollinated (ornithophilous) flowers.[5] But features can also be said to "coevolve" *within* species. First among these is the coevolution of life's somatic and reproductive features and their two broad and indispensable properties: survival and procreation.

In all this, natural selection acts on the features of the somatic embodiment of individual organisms, but that is all it does. Its effect on reproduction may be important, but it is always indirect. Its primary role is as the agency of selection for the somatic features of life. Its effect on the evolution of the mechanisms of reproduction is secondary.

As a result, when natural selection is considered, the number of offspring or fecundity is not the only measure of reproductive success. More is not necessarily better when comparing species. As I explained, the single issue of humans and the small number of offspring of other mammals may reflect a more successful reproductive process than a plant or insect that produces enormous numbers of offspring. The size of the brood reflects the attempt to produce enough progeny to guarantee the continued survival of the species given the challenges its offspring face. Some species have only a few offspring, but there are enough for it to persist, even thrive. On the other hand, a species that produces a great many offspring may fail as its progeny are decimated by life's circumstances, by natural selection. It is the balance between reproduction and survival that is critical.

Chapter 17

THE FINAL CAUSE OF REPRODUCTION

What Caused Reproductive Evolution?

Well, here's another nice mess you've gotten me into.

Oliver Hardy to Stan Laurel in many films

Having outlined the means of reproductive evolution, we can now turn our attention to the *final cause of its purposes*, the reason for its existence. To this point we have learned that natural selection serves as the bridge between the final cause of the somatic features of life (environmental danger) and their purpose (survival).[1] We have also learned that a reproductive selection links the purpose of reproductive traits (life's continuance and evolution) to a final cause that we have yet to identify. In trying to identify it, we ask a question that is at the same time frivolous and profound: "Why in God's (or Darwin's) name do we have children?"

WHERE TO LOOK

To the best of our knowledge the objects and forces of the physical world care not a whit about life, no less about the creation of new life or the evolution of the mechanisms that produce it.[2] Surely, this is not a good place to look for the final cause of reproduction, for the final cause of the creation of new life.

And outside of self-serving actions, living things themselves show much the same disinterest. The coldness most organisms display to others, whether we are talking of aggression, like cheetahs on the hunt, or simple indifference, gazelles grazing or crushing bugs underfoot, appears to extend to members of their own species, even intimate coevals (think of plants and invertebrates). And putting humans aside, they show no interest in abstractions, even one as important as the continued existence of their own kind, let alone living things in general.

Except for occasional, sometimes extraordinary examples of caring, life, as Darwin explained, is pretty much a self-centered business. But then why the oddity, the enormous idiosyncrasy of reproduction—a process centered on others, not ourselves—with the grand purpose of ensuring not our survival but that of life, along with its evolution. Without reproduction, life in all its forms would end in the blink of an eye, in a mere matter of years. And who would give a hoot? Yet someone or something appears to. Despite their otherwise selfish nature, that someone or something seems to be the living things themselves.

Here are two powerful examples of the *selfless* interest of living things in creating new life. We have talked about the first—sexual attraction. It and the ensuing impulse to reproduce are sometimes so powerful, so commanding, so utterly gripping that they can overwhelm the urgency to survive. Though it often masquerades as a self-seeking act, sex can be undertaken with such reckless abandon, overweening libido, and even rapture that reproduction takes place not only when it provides no selective advantage to the participants, but also when it is downright dangerous and, for some, certain to end in death. Such obedience to the imperative of reproduction, ensuring fulfillment of *its* purposes, suggests that interest in the continuance of life, at least as it relates to one's own group, is abiding.

The second example is even more compelling. I am referring to certain aspects of maternal altruism, the concern of the female for her offspring. This is not the altruism of love and compassion that we are used to thinking of, that we engage in as a matter of choice. There is no choice here, no possibility of equivocation, nor of a hidden self-serving agenda, but neither is there devotion or kindness. In its terms, the female cannot be an uncaring mother. She is duty bound. It is altruism in its *pure* form—rigorous, unambiguous, mechanical, and determined. The female's actions are innately and indisputably altruistic.

Though the traits are well known to us, their stunning implication has not been appreciated. For instance, the female mammal goes to enormous lengths to welcome the fetus. She provides a nurturing environment for its development and birth, in everything from the implantation of the embryo, to the development of a placental circulation, to uterine contractions, all of it in the right place at the right time in order to meet the needs not of the mother but of the fetus, of the developing embryo.

Or, after birth, the mother's mammary glands produce milk, again not for herself, but to serve the suckling child. The amount she produces is matched to the needs of the children of her species—their size and the number in the brood. A mother may decide to reject a particular child, refuse to feed it, or for that matter refuse to feed them all, but whatever her behavior, evolution has equipped her with this other-directed apparatus, this "altruistic machinery." We are called "mammals" not just because females have mammary glands, but because the purpose of those glands is feeding the young.

Similarly, for egg bearers, egg coverings, eggshells, and the like protect the fetus, while at the same time the egg's contents provide the nutrients needed to sustain it in its development. Again, though the egg is the product of the mother, it is made not for her but for her offspring. As in the mammal, the mother's efforts are tailored to the needs of the fetus. The thickness of an egg's shell and the amount of food it contains correspond to the requirements of the developing organism. Shells may be thinner or thicker, or there may be only a soft covering, and eggs may contain more or less food based on the severity of the environment the fertilized egg is likely to encounter, the duration of this phase of the life cycle, and the size of the progeny. Even when the mother abandons her offspring after laying her eggs, their existence is evidence of her concern for her offspring, however involuntary. Evolution has commandeered her resources for *their* benefit.

THE FINAL CAUSE

These compulsions make it clear that we should look for reproduction's final cause *within* living things. At least for our own insular reproductive purposes, we seem to be its authors. Yet there is an enormous difficulty with this conclusion. There are two overarching problems that, taken together, are devastating. First, we cannot look to reproduction itself as the final cause. As explained, it cannot be its own cause. But likewise, and here is the rub, we cannot look to our somatic features either. Whether for individual traits or for all traits combined, our somatic embodiment is concerned with survival and not reproduction. So if reproduction cannot be its own cause and if our somatic incarnation has no interest in it, then we have excluded the whole of life. We have ruled out all possible sources within the organism as the source of reproduction's final cause.

You might wonder about sexual attraction. After all, it leads to sex, and sex leads to progeny. But it can be the final cause of reproduction no more than fertilization can or the birth of a newborn. As part of the reproductive system, it cannot cause the whole. It is merely one step, albeit the critical first step, in the process. It cannot be the final cause of reproduction any more than a steering wheel or engine can be the cause of a car.

But if reproduction's final cause is not found outside or within us, then where is it found? We seem to have ruled out all of the options in the material world. Even so, we do not need to evoke metaphysics to explain the evolution of reproduction. There is one option left and it is undeniably a denizen of this world, though it is neither outside us nor within us. Then where is it? What is it? We can identify it by asking, "If survival maps onto danger for life's somatic features, then what do the purposes of reproduction map onto?" What is reproduction's "danger"? Unquestionably, it is not the same danger that helps organisms survive life's depredations, since the purpose of reproduction is to produce replacements for those who fall prey to them.

Well then, what is reproduction's "danger" and where is it found? Amazingly, it can be unearthed in the ultimate or final engagement of living things with the environment. It is found in *death*. Death is neither within the living body nor outside it, but needless to say it is of this world. It is, so to speak, at the *boundary* between the living object and the environment. In the same way that danger is the final cause of life's somatic purposes, death is the final cause of its reproductive purposes. If means of survival is the consequence of danger to survival, then the means of reproduction, the agency of life's continuance, is the consequence of the need created by death.

Imagine a circumstance in which there is no danger. As discussed, in this case there would be no need for the adaptive mechanisms that characterize life's somatic embodiment. There would no danger for them to protect against. Likewise in the absence of death, there would be no need for new beings, no need for reproduction and its mechanisms. Life's continuance would be ensured by already existing life. Such a world, a world without danger or death, would be an everlasting Garden of Eden, a place in which mechanisms of adaptation and reproduction are unnecessary. But, as the Hebrew bible tells us, such a place could only have come into being by God's hand. Without such a dwelling, life is accompanied by danger and death, by adaptations and reproduction. It cannot exist without them.

Still how can death be a final cause? How can it have given rise to the mechanisms of reproduction in the same way that the somatic adaptations of life evolved as the result of environmental danger? The danger to life presented by the environment is dynamic (or kinetic), the result of the actions of material bodies. It is in their actions that they pose a threat, that they embody danger. We can point, one by one, to the specific threats their multitudinous actions cause and can analyze their nature, often at a very deep level. However, in the absence of action, material existence is benign.

It is like death. Death is not dynamic. There is no action. It is ceaselessly and forever static. A dead object is inert. It merely exists and as such is benign. Though it may be acted upon, it does not act itself. It cannot cause anything in its own right, no less serve as a final cause. A rock may fall and cause a response, but sitting on the ground, it causes nothing unless you trip over it. A dead body is like a sedentary rock. It cannot act of its own accord.

But—and this is critical—this is not the death I am referring to. The only impact the death of an *individual organism* has on life's continuance is negative. There is one less living thing and one more inert object. Nor am I talking about the extermination of lineages. Obviously death can be of no value to groups that no longer exist. What I *am* speaking about are groups that are still flourishing, *when viewed as a whole*, as a group, however individual members fare, independent of their fate.

It is in this context, in the context of the group, that death serves as the final cause of reproduction's purpose, as the final cause of life's continuance.

Curiously, when considered in these terms, death takes on a dynamic character. For the group overall, for the population of organisms in its totality, it is not the death of individual members that matters, not the absolute number of failures, not the *number* of deaths, but the *percentage* of those born that are culled before they can reproduce. As part of a population, as an element in a percentage, the moribund becomes dynamic and the dead object becomes the final cause that calls for life's replacement. How absolutely astounding—the dynamics of populations give a dead being a dynamic quality. Even more remarkably, the grip of sexual attraction and the mandatory altruism of the female evolved not to serve *our* desire to have children, our intimate needs, but to help ensure the continuance of our reproducing group, our species, the population of organisms of which we are but an impersonal part.

DARWINIAN THEORY AND REPRODUCTION

In the face of unrequited danger and unanswered death, life ends. In this, evolution presents two ironies. The first is that the *danger of death* leads to the evolution of the somatic features of *life*. It leads to life and its adaptations. The second is that *death* (not the danger of it) leads to the evolution of the mechanisms of reproduction, to the means of life's *creation*. This is confounding. It contradicts common sense. For instance, death is doubtless a result, not a cause. Indeed, it is the final result of life. But as it turns out it is also a cause. It is a result for the individual and a cause of action for the group.

The evolution of life's reproductive features, what death gives rise to, differs from the rest of life's embodiment, that concerned with survival, in the following essential ways:

- First, foremost, and most obviously, the means of reproductive evolution concerns the creation of new beings, not the survival of extant organisms.
- Reproductive selection does not take place for the individual features of the reproductive system but for the system as a whole.
- The reproductive system of individual organisms is not the subject

of selection, but selection occurs among groups of organisms, among reproductive communities.

- The group's fate does not depend on what happens to individuals in their reproductive endeavors, but rather the ultimate fate of the individual's patrimony depends on what happens to the group.
- Selection does not occur during the actions of its features, that is, during reproduction, but after their task has been completed or before it has begun.

And finally,

- Reproduction and its evolution are not concerned with life's circumstances in the present but with an unknown and uncertain future.

Of course these are neither the means nor the mode of natural selection. And yet, as we have learned, natural selection nonetheless plays a prominent, albeit indirect and secondary, role in reproductive evolution. Along with the number of progeny, it determines how many survive life's tests and reach sexual maturity to reproduce. Likewise, it plays a role in the attempt of the somatic and reproductive features of life to coexist in one body, to harmonize. In the final analysis and in line with Darwinian theory, natural selection is critical to the evolution of both the reproductive and somatic features of life, though in regard to reproduction, its role is secondary to a primary reproductive selection. Moreover, even when natural selection is involved, the criterion of merit for reproductive evolution is no longer extant danger but death.

The final causes of somatic and reproductive evolution, danger and death, are connected to each other in another broader and even more important way. Like the somatic features of life, the evolution of its reproductive traits is a mechanical product of the material world; its purposes and final causes arise from it. In this, somatic and reproductive evolution are strictly analogous. Whether the measure is quantitative, as with the number of progeny, or qualitative, as with survival, there is no need for a supernumerary element.

Nature took its ordinary, intrinsic course in the evolution of both. In the same mindless way that it cares about life's survival, it cares about reproduction, about life's future.

Chapter 18

"FOR THE SAKE OF WHICH . . ."

The main Business of natural Philosophy is to argue from Phaenomena without feigning Hypotheses, and to deduce Causes from Effects, till we come to the very first Cause, which certainly is not mechanical . . .

Isaac Newton, *The Principia: Mathematical Principles of Natural Philosophy*, 1687

Aristotle said that there are two kinds of final causes: that "for the sake *of* which" and that "for the sake *for* which." The first simply does what it does. It gives rise to purposes merely as a consequence of its existence. A purpose may or may not be beneficial, but if it is, the benefit is not premeditated, it is not so by design or plan. The second type of final cause intends just that. By thought or action it is calculated to produce a benefit. It plans what it does and for this reason is commonly associated with God or a God-like entity.

We have been talking about purposes that arise from the first type of final cause. For the somatic and reproductive features of life, the causes are danger and death, respectively. Darwin's theory of natural selection is based on danger, while evolutionary or population genetics or dynamics, the child of the modern synthesis, is based on death. Neither the existence of danger nor the occurrence of death provides a plan or benefit. They merely produce a result that fulfills a purpose.

LIFE WITHOUT DESIGN

William Paley, the philosopher, theologian, and prior resident of Darwin's rooms at Cambridge University, tells us in his 1802 book *Natural Theology* about a walk across a heath during which a human observer makes two comparisons, one between a stone and a watch and another between the watch and a living creature. He sees the watch on the ground alongside the stone. The stone, he surmises, has probably been lying there doing nothing, inert, without means or purpose, whereas the watch is mechanical and has the purpose of keeping time.

As for the living creature, following the ideas of Descartes and others, Paley thought of it as a complex machine that was like the watch except that it was designed and manufactured according to God's, not man's, blueprint. In this comparison Paley was making a teleological argument, an argument from purpose, for God. Humans are the final cause of watches, the God of watches. Whereas the final cause of humans and all living things is God almighty. As man designed watches, so God designed life.

Darwin believed that his theory of natural selection was a refutation of Paley's watchmaker analogy. Life, he argued, was wrought by natural selection, and that as such evolved without design (even if with purpose), Godly or otherwise. The theory was a repudiation of the second form of Aristotle's final cause as the source of life, the cause that intends to provide a benefit, but was perfectly consistent with the first, a final cause without a plan.

ON THE NATURE OF SELECTION

In his 1986 book *The Blind Watchmaker*, evolutionary biologist Richard Dawkins envisions natural selection as "a blind watchmaker," an unsighted agency that is ignorant of what it created, as with the first kind of final cause. But this was not his meaning. He was not envisioning a final cause. His use of the term was merely a literary trope. He was not making a teleological statement about the relation between watchmakers and natural selection. He was *not* saying that as a blind watchmaker natural selection was a blind final cause.

As we have discussed, natural selection is not a cause at all. It causes nothing; it makes nothing. It is not a watchmaker, blind or otherwise. It is not even a mechanism. According to Darwinian theory what it does, all it does, all nature does, without thought or intention, is *choose among diverse objects that already exist*. The term *natural selection* just describes the automatic *choices* of nature that are ultimately responsible for the evolution of life's somatic features.

We have talked about how selection makes something that is inordinately unlikely, such as the spontaneous appearance of life, achievable with surprising rapidity. The example I gave was from *The Blind Watchmaker*. Dawkins used a simple sentence to show how this works. Selection was made on the basis of the goodness of fit of randomly chosen sequences of letters to the chosen sentence.

However effective, this selective criterion, the goodness of fit of the letter to the chosen sentence, was not really his choice to make. As a matter of fact, there really was no choice in the criterion used. For the sentence to emerge, no other measure—say, the frequency of various letters in the language or the number of straight lines or curves in different letters—would do. Only the goodness of fit of the letter to the chosen sentence would work. This is true of selection in general. Not only is the success of a particular criterion of selection determined by its *objective*; nature only provides a single criterion of selection for each objective of selection. The two are related as obligatory pairs.

Say that natural selection was based on the number of spots on organisms. In that case, the only evolution that could take place would be in relation to "spottedness." For life's somatic incarnation and its adaptations to evolve, the criterion of selection *had* to be the relative fitness of objects to their environmental circumstances. There was no other choice, no other option. The somatic features of life evolved to survive the press of danger. Similarly, selection for the evolution of the reproductive mechanisms of life had to be based on the number of progeny that were produced and reached sexual maturity. Their number had to be sufficient for life to continue despite the occasion of death. No other assessment would do.

And as it turns out, these criteria—fitness and the number of progeny—

cannot be attributed to a reductionist biology in which chemical reactions and physical states explain all that needs explaining about life and its evolution, because neither danger nor death can be reduced to them. Though both danger and death arise from events in the material world, in their own right they are the substanceless causes of life's survival and continued existence, of life's evolution. Death even leaves the master chemical DNA and its code bereft—present but lifeless. It is in danger and death that we find the blind watchmaker. Danger drove the evolution of life's somatic features and death that of its reproductive features. Both are blind and both are causes of life's evolution.

REPRODUCTION AND TRANSCENDENCE

Though neither benevolent nor revealing a greater plan for life's continuance, the imperative to reproduce is truly remarkable. It goes beyond the mechanisms of reproduction and their evolution, beyond their mechanical and automatic properties, indeed beyond their material existence. Astonishingly this product of death and a simple counting of numbers displays an interest in posterity outside the limits of experience.

Consider its goals:

- to produce new life, create new organisms;
- to secure the continuation of reproductive groups as entities unto themselves;
- to ensure the posterity of one's own family line; and finally
- to guarantee the future of life itself.

Each of these aspirations lay beyond the here and now, and yet the features that fulfill them evolved as a reflection of the number of offspring and nothing more. The goals listed are of course totally unlike those that underlie natural selection. But there is an even more fundamental difference between them. For its part, natural selection is not merely a biological phenomenon. It is a feature of the broader material world that is as basic as light or diffusion. Whatever life's fate, natural selection continues cataloging the disparate destinies of contending physical and chemical entities.

Reproductive selection, on the other hand, ends when life ends. Although it is a basic fact of life, it is not a fact of nature beyond living things. Its restriction to living things means that reproductive selection is *less* fundamental than natural selection. Yet, at the same time, because its choices are limited to living things, it cannot be reduced to or fully understood in terms of the inanimate properties of the material world. As such, it lies beyond or transcends them.

Still, despite this break with the material world, reproduction's interest in posterity may seem narrow, restricted to one's own group, indeed restricted to the nearest, the adjoining generation and the ability or inability of the current generation to replace its kind. While this is mechanically true, reproduction's interest in life is not nearly so circumscribed. Indeed, quite to the contrary, it is all-inclusive and never ending; its absorption with life, while parochial, is also catholic.

In the first place, there is the slight if barely noticeable impact of the reproductive fate of even the narrowest of family lineages on each and every larger group to which it belongs, from species to the whole community of living things, and not just in a given generation but for all generations to come. But reproduction is all-inclusive and never ending in a far more imposing way.

Obviously, whatever the means and circumstances, all living things reproduce. Needless to say, to continue to exist, they must. And so the reach of reproduction, its concern for the future, its transcendent character is at one and the same time as narrow as a family line and as broad as all of life, encompassing as it does each and every species. Just as all living things are concerned with their own survival in the present, reproduction is concerned with the future of life. Through it, life's interest in its own continuance extends throughout the biosphere, present and future, taking in the entire world of living things and casting its great immaterial shadow on generations to come, on those that do not yet exist, on those not yet conceived, on those unknown, unspecified, and uncertain.

Chapter 19

ON CHAIRS AND CAROB TREES

Human speech is like a cracked kettle on which we bang out tunes that make bears dance, when what we want is to move the stars to pity.

Gustave Flaubert, *Madame Bovary*, 1886

Though numbers of offspring are about as far as one can get from a final cause with a plan in mind, the mechanical nature of the causes that invest the products of both somatic and reproductive evolution with meaning does not allow us to conclude that deliberate or premeditated causes, causes by design, do not exist or are limited to metaphysics, or even that life as we know it can be fully accounted for in their absence. To the contrary, whatever your proclivities, the conclusion simply cannot be avoided that such causes *are* found in the material world, most notably in human intentions and inventions.

There is a great irony in this. It was the existence of human invention and planned action that provided the basis for great philosophers such as Plato and Aristotle, for the Hebrew sages and countless other thinkers, as well as for ordinary people, to imagine a god (or gods, as the Greeks and Hindus had it) that had the same, though an incomparably greater capability for invention and creation. *Like humans*, gods could be final causes. Without human invention we might ask, would we have been able to conceptualize a creative god?

THE GOD OF CHAIRS AND CAROB TREES

Here are two examples of *humans* as final causes of invention and creation, as final causes that *for the sake for which*, that are taken from our previous discussion. The first is about chairs (discussed in chapter 14) and the second is in the epigraph at the beginning of the book from the Talmud.

I have described the final cause of chairs as gravity, a cause of the first kind. But of course gravity never imagined chairs, let alone built them. Chairs are undoubtedly an invention of the human mind. As much as gravity, a human with a plan for its embodiment was its final cause. So we can say that chairs have two final causes: one without a plan, gravity, and another with a plan to provide a benefit, a human inventor. This is true for human artifacts in general. For instance, composers have all kinds of external reasons for writing music, everything from sex to war, from hope to despair regarding their personal situation or the human condition, in search of life or fear of death, for love of beauty, or simply for money. There is always a reason for creating music that is external, drawn from the environment, a final cause of the first type. Yet no external cause has ever written a note of music. That of course comes note by note, bar by bar, melody by melody, and rhythm by rhythm from the mind of a human composer, its inventor and the other final cause.

FOR THE BENEFIT OF POSTERITY

As for the epigraph from the Talmud, a minor Hebrew sage, Honi, tells about coming across a man planting a carob tree. When asked how long it will take for the tree to bear fruit, the man replies, seventy years. Honi inquires if the man expects to be alive in seventy years, implying that this is not likely. Why then, he asks, plant the tree when he will not benefit from its fruit? The man answers that he found trees that his forebears had planted for him, and he is doing the same for his children. In this he is not referring just to his actual children (given life expectancy at the time in all likelihood they would have died before the first fruiting) but more generally to posterity.

The man in the story is the final cause of the tree's subsequent life. His thought and action set the course for its future existence. Whether we are talking about the man planting the carob tree, the carpenter with the chair, or the composer with a melody, a human conceived them all and did so with a benefit in mind. They are all final causes of the second kind. There is, however, an important distinction to be made between the man with the carob tree and the inventors of chairs and music. His labors were meant not for his own benefit or even for the benefit of those around him but, like reproduction, for an abstract posterity.

When Honi asked, why labor for the benefit of some unknown person in an uncertain future, he was talking about a concern for humanity based on Hebrew religious teachings. But outside of a religious commandment, what could have been the source of this speculative benevolence? We have no difficulty understanding the human motivation to invent and build for self-interested purposes, but why create for others, especially those whose existence is merely imagined or a matter for speculation? After all, his tree may have grown to fruit with no humans in sight.

This brings us back to reproduction's interest in posterity and the question: Why in God's (or Darwin's) name do we have children? In asking this question I used a common figure of speech, "Why in God's . . . name," as an intensifier that roughly means "Why on earth?" It specifies feelings of regret or even horror at one's actions at having children. The statement is plural to signify the universality of the thought, and Darwin's name is added in parentheses because the question being asked is not about God's purposes but about the seemingly odd, yet all but ubiquitous biological imperative of having children.

Unlike the man and the carob tree, reproduction's interest in posterity is not one of benevolence. Its final cause, death, is of the first or unintended kind. The world seems to hold no plan to benefit living things by ensuring the continued existence of their kind. The agency of evolution for the reproductive features of life, based on the number of offspring and death, evinces neither a plan nor a benefit. Whatever psychological and sociological reasons there may be for the female's desire to have children and the male's desire to father them, all species are driven to procreate by inherent wants born of reproductive evolution. Despite the pain and beauty of human sexual longing, despite sensual and romantic feelings, despite the wonderful words of our poets about love, in this sense our reproductive desires as well as the transcendence of the process seem no different at base from those of worms.

We may derive great pleasure and happiness from them and from what we have created around our sexual urges, and though children are a magnificent achievement worthy of celebration, when looked at with the cold, clear eye of the biological imperative to reproduce, our sexual needs seem to be mere contrivances that encourage us to do death's bidding. Death is the ultimate source of our desire, and there is no benevolence in that. And yet it is death that leads to reproduction, and it is reproduction that transcends the material world as it expresses life's great yearning to persist.

There is something else. The irreducible unit of selection for the evolution of the somatic features of life is the individual living being, whereas, however temporary, the irreducible unit of selection for the evolution of its reproductive features is the family. In humans and some other species, it is in the family, in the reproductive unit that benevolence toward others arises. Juxtaposed against the self-serving interests of the individual honed by natural selection, the parents impose order to benefit the whole family, to benefit others. Notice that the man in the Talmud story refers to his children even if the reference is to generations yet to be conceived.

In modern times, particularly in developed societies, this imperative may be devalued, as the child rules supreme or the parents' narcissism trumps all. Nonetheless, the importance of shared effort and sacrifice is the basis of the family unit. And as the irreducible unit of reproductive selection, the family is not merely a social construction, any more than species are just social constructions. It is in the family unit that the foundation for a deep biological benevolence toward others is found, including benevolence toward the broader community that is given voice in the Hebrew commandments. There may be no benevolence in reproduction, but generosity toward others arises from it, from its family, from mother, father, and children.

IN PERPETUITY

It is appropriate to end this narrative with the most astounding of all facts about reproduction. While there would have been no life, no evolution without reproduction, *with it*, absent a large celestial collision or until the sun gets hotter on its way to becoming a cool red giant, life's presence on earth seems inextinguishable, eternal, despite the fact that the earth is no Arcadian paradise.

Indeed, it has been estimated that the powerful destructive forces of nature and its natural selection have destroyed over 99 percent of all the species that have ever existed. And we can confidently say that most of those flourishing around us today will eventually be annihilated (though some few, such as certain bacteria, seem to have persisted for a very long time). Despite the necessity of living things destroying each other to survive and the relentless destructive the forces of the physical environment, life has not only continued undeterred; it has progressed greatly in the over four billion years of its existence.

And as we look to the future, the expectation is that things will be much the same. Despite the continued extinction of species, new ones will emerge and thrive, life will persist and continue to evolve until it ends in a physical catastrophe or with the end of the solar system. Life's persistence in the past cannot be disputed, and its continued doggedness in the future seems a good bet, and yet all we can offer as the reason for this ostensible invincibility is that reproduction takes place. But reproduction is the miracle of perpetual existence only *if* there are creatures to reproduce, if life and reproduction are not ended by nature's unkind hand, by the ravages of existence. And nature offers no guarantee of this. The question then is, why, in the face of it all, all the depredation and destruction, has life endured, and why does it seem likely that it will continue to endure for a very long time to come, despite its constant, indeed never-ending failure?

It is said that once we are past the reproductive stage of life, or if we lack the opportunity or ability to have sex and reproduce, that we are of no use to life. We are excess baggage. If this is true, then our somatic means and mechanisms, our adaptations, do not exist in the service of *our* survival, as Darwin said and as common sense dictates. Though no doubt we survive because of them, their true master is posterity. Their task is to ensure reproduction and thereby guarantee life's continuation and evolution. From this point of view, it seems that our somatic and reproductive adaptations, the features that form our being, are not there to serve us but to serve life in its grand march.

Put another way, since our adaptations are what comprise us, our purpose, our raison d'être is life's continuation and evolution. We live to reproduce. Life has evolved to promote its own continuance and evolution, to make itself inextinguishable, and we are the vessels that serve this overarching task. Through our adaptations, our myriad struggles, strivings and exertions we reach beyond the world in which we live and function, to serve the world wrought by reproduction that is not only outside our personal experience but that astoundingly has yet to be created. We serve an unknown future. Unlike the trajectory of a thrown stone—knowable but latent—we face a future of prospect and potential, intangible and unknowable. Reproduction is about creation in the deepest, most abstract sense.[1]

CAPITULATION

Mark Twain is said to have complained that the difficulty he had with his short novel *The Tragedy of Pudd'nhead Wilson* was the loss of control *not* of his material but of his characters. They had come to have a life of their own that he just could not control. They wrote the story they wanted and were preventing him from writing the one he intended. Some of them so exercised him that he threw them down a well.[2]

I have similar feelings about *The Paradox of Evolution*, though my wrath is not directed at characters of my invention or even at the imposing ghosts of Darwin and Wallace, whose ideas were hardly an impediment. Though I have no characters to throw down a well, nonetheless, as I wrote the book, it took on a life of its own. I often felt as if I were sitting on a tornado, holding on for dear life, as my suppositions and biases were being tossed one way and then another, not by me but by nature.

Whatever my intentions, the book has turned out to be a murder mystery with the killer being none other than death itself. Both surprisingly and predictably, the cause of life's epochal persistence is death along with its boon companion, danger. They are the wellsprings of life's constant renewal.

CODA

As explained, *The Paradox of Evolution* is the third book I have written exploring the reductionist idea that we can fully understand the processes and mechanisms of life through deep and comprehensive knowledge of their constituent parts; their underlying structures, molecules, and reactions; and critically life's central chemical substances DNA and proteins. I have called this belief strong micro-reductionism—full (strong) understanding of living systems can come from deep (micro) knowledge of their parts.

There can be no doubt that the reductionist impulse has been responsible for our far-reaching mastery of life's material nature, for the great accomplishments of biochemistry, genetics, cellular ultrastructure, and much more. Yet the widespread and often heralded effort to apply knowledge of the parts of living things to capture the properties of the mechanisms and processes of life as they are found in nature, whole and intact, is not only fatally but fundamentally flawed. It can obscure rather than illuminate our understanding of life.

LESSONS FROM THE LIVING CELL

In the first book in the series, *Lessons from the Living Cell*,[1] I considered the application of this belief to the modern biological research laboratory. When understanding is sought by means of research undertaken from the strong micro-reductionist perspective, speculation and supposition often masquerade as true knowledge. This is due to a most important misunderstanding of the scientific method. In particular, science's central rule is not applied. To wit, to establish whether a scientific theory, model, or hypothesis is true, it must be tested against the properties of the mechanisms or processes it claims to explain as they are found in nature.

The strong micro-reductionist thinks that such tests are unnecessary and believes that the truth can be apprehended by means of weak inductive inferences—I have observed this, and this is what I think it means (or more forcefully, this is what it *does* mean). By this sleight of hand, the researcher's hypothesis becomes established fact. What is believed and what is true become indistinguishable. As a result, the truth becomes a matter of inclination or partiality, not proof, and scientific knowledge little more than the stuff of taste or dogma. However well informed, however overweening the opinion, opinion is all there is.

In fact, it is all there can be. Without testing our theories and hypotheses against life's mechanisms and processes as they are found in nature, whole and intact, there can be no honest, no forthright way of knowing whether or not they are correct. Even the claim that this or that part of an object is actually a component of a particular functioning system cannot be justified. Without the salve of good fortune, in the world of the strong micro-reductionist what is held up to be knowledge and understanding is no more than a guess or, worse yet, solipsistic ignorance or even duplicity.

Paramount to testing theories is the principle of *falsification*. Falsification is the rejection of a scientific theory in the face of its failure to account for properties of the natural system it is claimed to explain. In testing a theory, we ask whether things occur as it predicts. Quite simply, if they do not, then our theory is wrong; it is false.

The philosopher Karl Popper argued that all we could ever know with any certainty about nature is what we determine to be false.[2] Confirmatory tests are always provisional. This is because, however convincing they may be, they can never forswear future tests that will show the theory to be false. For this reason there are no definitive scientific understandings. However well established, however widely held, they can always be overturned, as Copernicus overturned Ptolemy and his epicycles, and as Einstein overturned Newton and his earthbound forces.

In *Lessons from the Living Cell*, I discuss two examples of the strong micro-reductionist impulse in biology.[3] One is about an explanatory theory for muscle contraction and is biochemical, while the other is about a theory that professes to explain how cells secrete proteins and is anatomical. In each case, tests against the properties of the intact system were thought unnecessary. Inferences (those of the weak inductive kind just noted) drawn from knowledge of the purported parts of the system (they could only be "purported" without knowledge of the intact system) were claimed to be sufficient to verify the theory, sufficient to conclude that it provided an accurate description of the workings of nature. But such a conclusion, however forcefully stated, simply could not be justified. Without testing the theory against the natural process, whole and intact, claims of verification, however appealing, however seductive are manifestly invalid. All there are, all there can be in this case are suppositions, presumptions, and untested guesses, smoke and mirrors dressed up in the guise of proof.

Though hardly new, we could even say "venerable," this way of doing research, rather than being excoriated by modern science, has had a resurgence. By stripping science of the premise that what we know is based on testing our theories against the properties of nature, what scientists believe to be true becomes a far more tenuous business, if not a different beast altogether. As physicist Thomas Kuhn explained in his classic book *The Structure of Scientific Revolutions*, in this circumstance scientific understanding is based not on tests of the truth or falsity of ideas but on shared beliefs. Those beliefs are usually based on evidence and proof but they are also based on all sorts of value judgments, preconceptions, suppositions, prejudices, assumptions, particular ways of reasoning and seeing, particular ways of making experimental observations, and finally on social and political relationships with their attendant hierarchies of control and power. Kuhn called these concoctions "paradigms," a comprehensive view of a particular aspect of nature that is agreed upon by a community of experts.[4]

Paradigms are the social contracts of science. They tell us what to believe and what to discount, what is possible, what is not, what is known and what is unknown, and how best to study it all. In this, the "truth" is no more than the common judgment of a group of scientists in a particular field. Certain evidence and interpretations are given the official stamp of approval, the imprimatur of scientific fact. Hovering over it all, usually denied but barely disguised, is advocacy, overriding political intentions of both a societal and a scientific nature. As for the former, the group tells us what society needs to do, for example, restrict carbon emissions; while for the latter, it tells us what science is to be believed, for example, global temperature is rising. Whatever the evidence and logic say, the truth is up for a vote among the cognoscenti, among the experts, a matter of their wants and desires.

From a methodological point of view, the central feature of the paradigm is that experiments and observations are designed to *affirm*, not test, its truths. Research seeks confirmation, not falsification. Indeed, falsification is not possible. A study or piece of evidence may fail to bolster the paradigmatic perspective, but it cannot show it to be false.

Any honest and knowledgeable observer of science, certainly of biology, would have a hard time disagreeing with Kuhn's view of its character. The picture he paints may be displeasing, we might wish it were different, but if we take off our rose-colored glasses and close our childish books about science, it cannot be denied. Despite self-serving pretensions, science is *not* simply or purely an activity in which humankind's hypotheses about nature are tested against its attributes. In Kuhn's view, it is little more than a system of beliefs, some justified, others unjustified.[5] When paradigms are overturned, as they fortunately are from time to time, whatever the rational basis for their rejection may be, the event is a political one, with human frailties, both admirable and contemptible, on prominent display. Whatever the facts, what was once held to be true is now discarded for no reason greater than that the beliefs of the community have changed. A change of heart has taken place.

But if this really is the way science works, how confident can we be that the principles we believe in and teach students provide an accurate account of nature's actual properties? In this regard, Kuhn thought that textbooks were an abomination. They taught unwary students that embedded beliefs were established facts of nature, when they were nothing of the sort.[6] To me the most damning thing about the scientific paradigm is that its beliefs are often fervently held in the absence of proper tests against the actual properties of nature. Science is deemed to be just fine, rigorous and explanatory without them.

LIFE BEYOND MOLECULES AND GENES

Underlying this strong micro-reductionist belief is the conviction that life's material nature is identical to, that is, coextensive with life itself. And this in turn means that our physical incarnation, our material embodiment is both a necessary and sufficient property for life's expression. What we are made of fully accounts for the phenomenon of life.

In *Lessons from the Living Cell* I explain that this is not true for the biological cell. Even this basic unit of life is more than the mere sum of its component parts, its molecules, their reactions and physical states, and the anatomical structures they form. Whatever the cell type, whatever its character, something beyond its material instantiation makes it alive.[7] In *Life beyond Molecules and Genes* I extend this argument to all the complex tissues, organs, and organ systems that comprise intricate metazoan organisms like us. Beyond this, the book is a search for what is missing, for that something else, that something more of life. Toward that end, it asks biology's age-old question: "What makes particular material objects alive?"[8]

Many biologists today think that this question is either irrelevant or worthless. It is irrelevant because modern science has exposed the chemical and physical glory that is life in its material instantiation. Life and the matter of which we are comprised (as well as its actions) is one and the same thing. Biology's central question, however sensible in the past, has become irrelevant. A particular material object is alive simply because it is that object.

But we cannot avoid the fact that this is no more than a statement of doctrine, of revealed truth, not science. The identity is claimed, not proven. And science does not, or at least should not, accept assertions as facts, however feverishly believed or however seemingly obvious. The proposition of identity must be demonstrated, specified, not merely declared. To do this we must ask and answer the following: "In what way does life's material incarnation gives rise to the living state?" What transition takes place as we move from the inanimate to the animate realm? What bridge is crossed, what chasm forged? What exactly makes life's material embodiment alive?

To answer these questions, to prove the identity, it must be shown that the material embodiment of living things is both a necessary and sufficient condition for the living state. If it is not, then the premise of identity cannot be true. There must be something else, something more that makes the particular material object alive. That the first predicate is true, that life's material embodiment is necessary, is obvious. After all, how can there be a living being without it? But the second predicate, sufficiency, is another matter. If life's physical being were also sufficient then, we would expect that *whenever* it is present so, necessarily, is the living state. The state of being alive would be an automatic consequence of the presence of our physical embodiment.

But even our most primitive human ancestors knew that this was not true. With death, life's material embodiment lingers, cold and inert, awaiting eventual decay and dispersion, while life itself is gone instantaneously. It is in this separation of life and its material incarnation that we find the basis of the soul. With death, it is believed, the soul escapes the confines of its material prison. But soul or not, life's end is accompanied by the loss of its *sufficient* property, whatever it is, while its physical embodiment remains in place to tell us that they are separate things.

To those who believe that the quintessential question about life's nature is not merely irrelevant but worthless, that the question has no answer and to look for one is a fool's game. For them, though our material nature can be spelled out in great detail, with impressive depth, specificity, and accuracy, aliveness simply cannot be defined. It is a vague and nebulous concept that will forever elude definition. Simply put, it has none, at least none that is not fatuous.

But how can this be true? After all, being alive is not an epiphenomenon but an authentic phenomenon of nature, or at least so it seems. Some things are alive, while others are not. Some are animate, while others are inanimate. Though some thinkers have tried very hard, this difference just cannot be disregarded or elided. If being alive is a true phenomenon, then it must be explicable even if we cannot explain it, even if we are ignorant of its definition.

Unlike those who do not believe that life can be defined, many today think that the enduring question about life's nature has been answered. Though not identical to its material incarnation, modern science has determined that life arises directly from certain of its parts. Indeed, unearthing the path from these parts to the whole living thing, unearthing a chemical path to life, is almost universally thought of as the most important accomplishment of modern biology, if not of modern science. It provides, or is thought to provide, an understanding of the living state that is as rigorous and deep as our understanding of the atom. Life can be reduced to just two types of molecule, DNA and protein. Put simply, DNA begets protein and protein begets life.

The problem is that this leaves us still unable to say what it means to be alive. Critically, neither DNA replication nor protein synthesis, nor both together are sufficient causes of life. Neither their presence nor that of their products guarantees life's presence. In fact, both replication and synthesis can take place in test tubes, in vitro, in the laboratory, in the complete absence of life. Living things do not emerge from these activities or their products.

Some have attempted to answer the question of life's ultimate nature by reaching beyond a simple material understanding. But they, too, have invariably failed. Indeed, their ideas suffer from the same shortcoming—the absence of life's sufficient property. If, as some think, life is an expression of complexity or can be defined as a formal system or as an energy-generating process, where in these concepts do we find life's sufficient cause? Why does this or that particular attribute make some material object alive? If we cannot specify why or, worse yet, if we can imagine objects with the same properties that are certifiably *inanimate*, if we cannot distinguish the animate from the inanimate, then whatever the concept, however fascinating otherwise, it does not explain why particular objects are alive.

If there is a distinction between the animate and inanimate, and the phenomenon we call life is not an illusion or delusion, it must have a definition. Not only that, but all we have to do to discover it is to identify life's sufficient cause. In *Life beyond Molecules and Genes* I offer a definition that accounts for life's elusive sufficient property. Life, I argue, is found in Darwin's theory of evolution by means of natural selection and its concept of adaptations. Adaptations are usually thought of as properties of living things, as features that help them survive environmental challenge. They are the evolved products of natural selection's choices, and as such reflect the relative fitness of organisms to survive their circumstances. But *Life beyond Molecules and Genes* says that this has it completely backward. Adaptations are not salutary properties of otherwise living things. They are not properties of living things at all. They are far more than that. They are what makes them alive.[9]

Adaptations, the argument goes, are life's source, its sufficient cause! As the book's subtitle instructs, it is "our adaptations [that] make us alive." Life attends their presence and disappears when they disappear. Life began not with reproduction or metabolism but with the first adaptation, the first property that provided a selective advantage to some ancient biological object in its struggle to survive. We can say that the number and character of adaptations determine an organism's degree of aliveness, and though we can survive without many, without others life cannot be sustained. Either way, with death each and every adaptation vanishes instantly.

What's more, the existence of adaptations and their central role in causing certain objects to be alive belies the strong micro-reductionist's claim that the whole is the mere sum of its material ingredients. Adaptations are not adaptive simply by dint of being, by dint of their material incarnation. Though their embodiment is obviously necessary, they serve an adaptive function only in the context of the living organism as it seeks to survive its engagement with the environment. It is only here, in the organism, undivided and unabridged, as it is found in nature interacting with its surroundings that life and its adaptive properties truly exist.

THE PARADOX OF EVOLUTION

This brings us to the current effort, *The Paradox of Evolution*. As the subtitle tells us, it is about the "strange relationship between natural selection and reproduction." Since the time that Darwin's theory first saw the light of day, natural selection has been thought to be the agency responsible for the evolution of all of life's features, including its reproductive features. Nonetheless, when reproduction is viewed in terms of its goodness of fit to Darwin's theory of natural selection, the two seem to be odd bedfellows. Namely, reproduction is about the generation of new life, while natural selection is about the maintenance of existing life. The former is about procreation, the latter about survival. One cannot help but be struck by the oddity of the idea that mechanisms that are about procreation evolved as the consequence of an agency whose only concern is the survival of existing creatures.

One aspect of this peculiar relationship directly concerns reductionism. It is the fact that despite the inclination to ascribe the evolution of all of life's features to natural selection, to this single common reductive process in which individual traits and their genes are subject to selection, the evolution of the reproductive features of life came about primarily by means of a distinctive reproductive selection in which individual features are *not* the subject of selection. In reproductive selection, the system as a whole, not its particular subsumed traits or genes, is judged.

There is another distinction between the two and it also seems peculiar. Unlike natural selection, reproductive selection does not occur during the germane events, that is, during reproduction. Unlike the emblematic struggle between the gazelle and the cheetah, reproductive selection takes place *after* reproduction is complete, as well as, even more remarkably, before it occurs. In any event, with the completion of reproduction, after the die is cast and settles, the only thing that can be assessed, that can be selected for or against is the *number* of offspring.

It is as a result of this fact that reproductive features evolved not individually but en masse, as an aligned group of properties. The assessment is global. Selection occurs for all the aspects of reproduction taken together. Individual traits are simply not considered. Though this is perfectly consistent with a reductionist perspective, there could not be a more dramatic rejection of its strong micro-reductionist form. Individual reproductive features in great part evolved as anonymous components in an overarching process.

Moreover, the evolutionary fate of an organism's reproductive system is bound to that of the group to which it belongs in a very different, indeed a contrary way, to that for the nonreproductive features of life. For the latter, what happens to individuals as a result of natural selection determines the evolutionary fate of the group, while for the former, the evolutionary fate of a particular individual's reproductive system is determined by what happens to the reproducing group or groups to which it belongs. Its fate is bound to the group's fecundity, to the success of *its* reproductive activities.

As a result of this dependence on the group, reproductive selection takes us out of the material world into a world of abstract mathematical concepts, in particular into the world of sets. Even when sets are collections of incarnate objects, they are not physical entities in their own right. They remain abstractions. And yet it is among sets or groups of living things, not individual organisms, that nature makes its reproductive choices. As bizarre as it may seem, in this case nature's ability to select among sets, between different abstractions, is neither mysterious nor miraculous. It is no more than a matter of counting. Though nature does not count things in the way that we understand counting, as a conscious activity, it nonetheless counts. Clusters within a reproducing group that produce more offspring are advantaged relative to those that are less fecund. In reproductive selection, as nature selects, it counts. As opposed to "as nature counts, it selects."

Beyond this, *The Paradox of Evolution* lays bare life's most basic nature. Though the adaptive mechanisms at life's center are enabled by their material embodiment, they are not mere incarnations of them. Neither adaptations nor the life they give rise to are simply expressions of their material nature. Though of course there can be no adaptations, indeed no life, without our material being, it is not the physical incarnation of our eyes, our heart, our skeletal muscle, our sinew, blood, and grizzle that is adaptive. It is not even our ability to run or fight, to pump blood or breathe. Features are adaptive not because of what they are or even what they do but *why* they do what they do. They are adaptive by reason of their *purpose*.

Nor are their purposes found in some sort of beneficent internal corporeal state. Quite the opposite; they exist in the context of perilous interactions between the organism and its environment. It is by means of these interactions that advantageous purposes, purposes that seek to protect us from harm, are fulfilled. In this, the trait's *purpose* makes it adaptive. If that purpose cannot be realized, then despite the continued presence of its physical incarnation, the adaptation disappears; it is no longer present.

Even though adaptations do not exist apart from the material world, in and of themselves they have no physical incarnation. They have no bits and pieces, no jib and sheet. They and their purposes are irreducible immaterial attributes that transcend the material world from which they arise. By extension, we can say the same about the property of aliveness. It is an irreducible immaterial quality that transcends the material world from which it arises.

Yet in spite of the seeming centrality of purpose to life, the concept has had an extremely difficult, not to say controversial history in biology. By the early twentieth century, as the reductionist perspective became dominant; most scientists came to reject the idea. Though it went without saying that biological mechanisms functioned, they did so without intention or purpose. This rejection of purpose was in great part the consequence of the understanding that purposes were God-given. Only God could invest our being with purpose, and since science sought to explain nature without invoking God, life's adaptive mechanisms could not have purposes however it appeared. And Darwin's theory seemed to explain how this came about.

As for reproduction, things were quite the opposite. It seemed to have a clear, indeed a self-evident purpose—having progeny. While we have learned (at least I hope that we have learned) that the adaptive mechanisms that are the product of Darwin's purposeless theory embody various purposes, as it turns out the self-evident purpose of reproduction is not a purpose at all. Reproduction cannot be its own purpose, any more than digestion can be the purpose of digestion or breathing the purpose of breathing. This does not mean that the reproductive processes of life lack purpose, any more than digestion or breathing lack purpose. They have a purpose, all right, but it is far grander than simply having progeny. Reproduction is not about itself, about having offspring, but about the seemingly altruistic purpose of the continuation of life.

Our somatic embodiment evolved to enable organisms to survive the challenges presented by the environment, while the mechanisms of reproduction evolved to enable life to continue. As said, the evolution of life's nonreproductive features is about the survival of individual organisms and is only indirectly about that of the group to which it belongs, whereas the evolution of life's reproductive features has it the other way around. It is about the continuance of *groups* of living things, up to and including the whole biota. The fate of the individual is not just secondary; it is immaterial.

In considering purpose in biology, we look to Aristotle for guidance. As he explained, events that have purposes necessarily have final causes. Something causes or acts to embody the purpose. This is often imagined in terms of a directing God, and this is the reason for the rejection of purpose as a matter of science. But Aristotle also envisioned a natural, as opposed to a metaphysical, final cause, one that lacks intention and acts without planning. It is here, in this natural causation that *The Paradox of Evolution* looks for the final cause of biological purpose.

In this light, biological systems have two purposes and two final causes. One pair of purpose and final cause is related to life's nonreproductive or somatic features, and the other to its reproductive traits. The purpose of the former is survival. All of our nonreproductive adaptations are in one way or another, in one fashion or another, to one degree or another devoted to our survival. The final cause of the purpose of survival is the *danger* that living things face in the world. Without danger, there would be no need for adaptive mechanisms. They would not exist, and for that matter neither would live things.

The purpose of reproduction is the continuance of the reproducing group, as well as all larger groups to which the individual belongs, up to and including the all-encompassing group of all living things. Its final cause is *death*. It is only with death that reproduction is needed to perpetuate life. So, the diverse adaptive features of living things are about two things: survival or continuance. In pursuit of them, they are driven by danger and death, respectively. The danger is to the survival of existing organisms, and it is death that prompts reproduction. Ultimately, neither is about the parts of living things. The causes of life's adaptive features, both somatic and reproductive, concern the fate of organisms and groups of organisms, not their parts.

TOWARD A GENERAL THEORY OF ORGANISM

Our story is almost complete, but one very important subject is missing. Though I have touched on it tangentially and discussed the centrality of DNA to modern ideas about life's nature, I have not to this point here or in any of my books given more than glancing consideration to what many would regard as reductionism's greatest achievement in biology—gene theory. With much justification they would maintain that along with cell and evolutionary theory it forms the basis of our modern understanding of living things.

The father of gene theory is of course the Silesian monk Gregor Mendel. His experiments in the middle of the nineteenth century and the meaning he deduced from them were not only seminal; they were models of micro-reductionist exploration. In his effort to explain inheritance, he reduced life to its components in two related ways. Every secondary-school student knows the first. He discovered that certain elements, *Elementen* in German, that we know as *genes* underlie and are responsible for the expressed features of living things, for their phenotype.

Though not forgotten, the second reduction has seemed trivial. To Mendel and the many scientists who followed, it was merely an experimental convenience. They broke life down, figuratively, into the various expressed traits they chose to study. Mendel examined seven aspects of the pea plant: seed shape and color, pod shape and color, plant height, and finally flower position and color. Of course, by so doing he left out many other features of the plant, some known to him, others not known to him or anyone at the time. But studied or not, together they were its *Merkmal*, its expressed characteristics. Whether in regard to the pea plant, or the inheritance of eye color or bone structure, and whether in flies or salamanders, Mendel and the geneticists who followed believed that what they chose to study were exemplars of all of life's features, all of its *Merkmal*.

This approach, choosing certain features to study, was not only successful; it was an absolute necessity if one hoped to discover what lay beneath the appearance of things. There was no other way to expose the genetic nature of life, to uncover its underlying elements. It is self-evident that nothing could be learned from the whole organism without reference to its specific characteristics, its distinctive components. Only then could life's genetic basis be discovered.

Life, Mendel had determined, was two things, *Elementen und Merkmal*, genes and expressed characteristics. Genes were the building blocks that underlie all the expressed aspects of living things, whatever the nature of the traits and whatever the species. Though his data was subsequently questioned by no less an authority than the great statistician Ronald Fisher, however accurate or inaccurate his numbers, Mendel was the first to demonstrate the existence of genes and their relationship to life's expressed characteristics.

Though Mendel's insights were pathbreaking, they left enormous gaps in our understanding. A striking and basic gap was that Mendel did not know what genes were. All he knew was what they did. Indeed, their material embodiment would remain a mystery for another hundred years. As explained and as anyone reading this most certainly knows, genes are thought to be sequences of subunits in the gigantic DNA molecule that provide a code for the structure of proteins. Proteins, the workhorses of life, are the most complex molecules known to us, and intricate metazoans like us, plants and animals alike, contain the code for some 20,000–40,000 different kinds.

Another enormous gap in Mendel's understanding was how genes actually give rise to the features of living things and thereby to the organism. Even though genes and proteins are understood today to be central actors in this process, how their digital world gives rise to the analogue world of the organism and its features remains an open question and an area of intense investigation even after all these years. In recent times efforts have focused on the regulatory mechanisms that control the time and place of gene expression. At what times and at what places within the materializing being are different genes expressed and particular proteins manufactured? The idea is that if we know the when and where of gene expression, the nature of embryonic development will be revealed.

Figuring all this out is not only a huge task with many factors to consider; it is an extraordinarily complex one. Some have suggested that even with much effort, unraveling it and its complexities will take the better part of the twenty-first century. Still, to many, however difficult the task may be, it is worthwhile. Given that we know the genetic code and that a comprehensive catalog of protein structure is no more than a matter of resources at this point, if we are able to develop a deep understanding of gene regulation, we will have captured life's three fundamental aspects: genes; proteins; and by means of gene regulation, the development of the organism from them. For all intents and purposes we will understand life in its entirety, from gene to organism.

Though the thought is seductive and the claim grand, the hope is false. Knowing how events are regulated is not the same as knowing the events. Though we may be able to determine when and where the conversion from the digital to the analog state takes place, with this information alone we would remain ignorant of even the most basic properties of the transformation. For instance, is what occurs commutative like addition, with parts (proteins and what they give rise to) being added to other parts to form the whole; or is it noncommutative, like chemical reactions in which new things are formed? And if both take place, what are the events and how are they connected?

The relationship between *Elementen* and *Merkmal* is thought to be much like that between atoms and molecules. Genes are to organisms what atoms are to molecules. Just as atoms give rise to molecules, genes, by means of proteins, give rise to life's features and thus to the organism. Whether or not the analogy is apt, the comparison highlights the enormous gap between our knowledge of how molecules are formed and how the organism and its features are formed.

The laws of chemistry are well worked out. They explain to a fare-thee-well how atoms give rise to molecules; that is, how the digital world of atoms gives rise to the analog world of molecules. Known "chemical bonds" provide the cement, and how they are created can be described with great kinetic and thermodynamic rigor. In comparison, we remain abysmally ignorant about how the features of living things arise from genes and proteins. We have no rules of organismal order and cohesion equivalent to those of chemistry.

There is one final gap in Mendel's theory and it is particularly thought-provoking. Though we can select this or that feature in this or that species to study, outside of their common physics and chemistry we cannot point to some general quality that applies to them all, to all *Merkmal* or for that matter to all organisms. All we can say is that genes provide the code for a *list* of protein molecules that in turn, as the result of some sort of subdivision or compilation, account for a larger *list* of enumerated features. All we have are lists: lists of genes, lists of proteins, lists of features, and of course lists of species. But however insightful the characterization may be, organisms are not lists of features any more than the attributes of those features are mere lists of parts.

All is divisible but not divided. Whether we are talking about traits or organisms, the objects are whole, entities unto themselves, integrated and unitary. Though gene theory is a universal theory in that it proposes that genes underlie all of life's features, we lack a comparable understanding of what they give rise to, an understanding of the organism and its traits that is as general and universal.

All we have are lists of diverse and sundry traits. Though we can spell out the differences and similarities in great detail, outside of physics and chemistry we cannot identify their common, their universal qualities, given that such qualities exist. Though we can match features or species based on shared appearances, composition, and purposes, we cannot recognize general attributes that are common to them *all* as we can with genes and of course chemical substances.

As the analog world of life comes into being, as its features emerge from genes and proteins, syntheses take place that by their very nature are not lists of things. Both the syntheses and their products would seem to have some common inherent property or series of properties of which we remain in great part ignorant. Given the existence of the syntheses and their products and our ignorance of them, Mendel's theory of inheritance can be said to be incomplete. It lacks not only a concept of the syntheses that take place but an understanding of what is being synthesized that is as general and declarative as gene theory.

It is not satisfactory to just say that the embodied organism is a coterie of features, and move on. Biologist Robert Rosen said critically that "(the modern view is that Mendel's) pea plant . . . is nothing but a bundle of discrete adjectives, which all together are equivalent to the noun they modify . . . actually identical with it."[10] But the pea plant is neither a bundle of adjectives nor for that matter a list of nouns. To stick with the metaphor, its features, the features of life, are sentences, often inordinately complex sentences that, taken together, form the narrative that is the organism. What is the common structure of these sentences and the narrative?

Expressed features do not merely sit side by side or even astride or tangled up with one another. Their associations transcend their individual nature. They are not a mere sum of separate things but are an integrated conforming whole. For instance, though the cardiopulmonary system of vertebrates designed to provide oxygen to our cells and clear away the end products of metabolism is the product of many genes and many proteins, and can easily be broken down into parts, it is not simply a composite of them. What comes into being during development, what evolved, what is synthesized is an assimilated or amalgamated whole that exhibits the manifold properties needed to carry out the tasks entrusted to it.

Though modern biology has a general theory of genes, of what lies beneath, it has no comparable theory of the observable features of life, or of the organism itself, or how it all comes into being. This leaves us in a most uncomfortable position. Despite the fact that we know a great deal about the parts of living things—their chemistry, physics, and genetics—and we likewise we know a great deal about populations of organisms, about their statistical character, inherited qualities, and of course how they live, sandwiched between them is the organism, life's central player, which, beyond cell theory, we only seem able to define, however ably, as a list of features.

But if organisms are indeed fully integrated wholes, not lists of parts, the result of a synthesis of some sort, without a general explanation for it and its subsumed features, the theory of inheritance is a half-baked cake, a story only half told. Even if we can say that genes give rise to the organism and its features, to life, we cannot say with equal precision, lucidity, and universality what as a general matter they give rise to beyond identifying names and lists of features. In *Life beyond Molecules and Genes*, I propose such a general theory. It posits that living things are not the sum of their parts but are an amalgamation of adaptive features.[11] Organisms do not *have* adaptations, but they are composites of life-imbuing adaptations. The living organism is the embodier of adaptations.

In this view, life's varied and sundry features are held together by their adaptive quality, by this singular, universal property of life. Still, even with this invasion from Darwinism, Mendel's theory of inheritance lacks an understanding of the organism and its expressed features of equal depth and clarity to the concept of genes. Given this and the other deficiencies mentioned, the search for understanding life's nature does not end, as the genome project some time ago so grandiosely proclaimed, with the discovery of the genetic code and the compilation of the "Book of Life" that the genome characterizes.[12]

FINAL COMMENT

In *Lessons from the Living Cell* we learned that a necessary preface to grasping the true nature of the mechanisms that underlie biological processes is knowledge of the properties of the intact mechanism.[13] Without it, we arrogate to ourselves knowledge that we simply do not have, knowledge that can only be found in nature. To study living things without acknowledging, understanding, and applying this essential truth leads to the pretension that our beliefs are correct, that we have uncovered nature's reality, when in truth all we have are the conjectures and suppositions of our imagination. Though in feigned humility we may admit that our knowledge is neither complete nor perfect, we nonetheless take our beliefs to reflect nature's truth when that designation has not been earned.

In *Life beyond Molecules and Genes* we learned that being alive is not a matter of matter, but of the relationships between organisms and their environment.[14] It is here that we find the sufficient properties of life. As such, what we call life or aliveness transcends its material basis and mechanisms. Life is a relational, not a material, property; it is about what occurs *between* different material bodies. It is found in our adaptations to the world we confront.

Finally, in *The Paradox of Evolution* we learn that life has two abiding purposes, survival—the work of the somatic mechanisms of the body—and the continuation of life—the work of reproduction. The final cause of the purpose of survival comes from outside, from existing danger, while that of life's generational continuation comes from death and the need to replace what is lost. In this, the evolution of life's somatic features is the result of natural selection's choices as life is lived, while that of its reproductive features is due to a reproductive selection that occurs in great part upon the completion of its creation.

None of these ideas is particularly complicated—scientific theories must be tested against nature; life is not found in its material parts but in the adaptations of the whole being to its environment, and while natural selection is about survival, there is a separate reproductive selection that is about the continuation of life. Nonetheless, as straightforward and in some ways as self-evident as these ideas are, they have in great part either been ignored or gone unappreciated. Though it is not clear why this has been the case, perhaps it is because acknowledging them would leave us unable to claim, as is the wont of our times, that the most vexing and important problems of biology lie behind us. That for all intents and purposes they have been solved by modern science. In this view, what remains, all that remains to be done, whether in the laboratory or the field, and however formidable, is the minor task of filling in the details, of straightening things up, of crossing the Ts and dotting the Is.

What reductionism has uncovered would probably leave even the most perceptive and forward-thinking Enlightenment figure flummoxed, totally disbelieving. And yet, despite these almost incomprehensible accomplishments, the end of biology is not yet nigh. Great questions linger, great secrets remain hidden, and great journeys of discovery are yet to be undertaken. Whatever our conceit, life is not yet understood in its fullness, in all its dimensions.

NOTES AND SELECTED READINGS

What has been written about Darwin's theory would fill a library. Outside of specialty scientific papers, news and magazine articles, several new books seem to appear every year that recount one or another aspect of the theory's history, focus on current issues, express a particular point of view, or all three combined. They include biographical accounts of Darwin's life, naturalist excursions that tell us about the remarkable diversity and beauty of life and its evolution or, as with today's Intelligent Design, present broad challenges to the theory's authority or defenses of it.

To my knowledge the most comprehensive catalog of references (over six hundred) is in Stephen Jay Gould's immense compendium *The Structure of Evolutionary Theory* (Cambridge, MA: Harvard University Press, 2002), and it hardly exhausts the subject. The references listed below include a small subset selected for their relevance to the contents of this book, and even these are not all-inclusive. Those directly related to evolutionary theory are either core to the subject or reflect my choices taken from a more extensive literature. Where possible I have avoided papers and monographs, primary and secondary sources intended for specialists and a technical audience, and where I have had a choice, I have focused on relatively recent publications. No doubt some of you will feel that I have left out one or another important citation. All I can do is apologize for the omission and promise to correct it given the opportunity in a later edition or in other writings.

PREFACE

1. Stephen Rothman, *Lessons from the Living Cell: The Limits of Reductionism* (New York: McGraw-Hill, 2002).

2. Stephen Rothman, *Life beyond Molecules and Genes: How Our Adaptations Make Us Alive* (Philadelphia: Templeton, 2009).

PROLOGUE

1. Helena Cronin, *The Ant and the Peacock: Altruism and Sexual Selection from Darwin to Today* (Cambridge: Cambridge University Press, 1991).

2. Charles Darwin, *The Descent of Man and Selection in Relation to Sex* London: John Murray, 1871).

3. The foundational ideas of the theory of evolution by means of natural selection are laid out in Charles Darwin, *On the Origin of Species by Means of Natural Selection* (London: John Murray, 1859); Charles Darwin and Alfred Russel Wallace, "On the Tendency of Species to Form Varieties and on the Perpetuation of Varieties and Species by Natural Means of Selection," *Proceedings of the Linnean Society* (1858) 3:45–62; Charles Darwin, *On the Various Contrivances by Which British and Foreign Orchids Are Fertilized by Insects* (London: John Murray, 1862); Charles Darwin, *The Descent of Man and Selection in Relation to Sex* (London: John Murray, 1871); Alfred Russel Wallace, *Contributions to the Theory of Natural Selection* (New York: Macmillan, 1870); and Alfred Russel Wallace, *Darwinism—An Exposition of the Theory of Natural Selection, with Some of Its Applications* (New York: Macmillan, 1889). Though no doubt a little stuffy by today's standards they are eminently readable even by nonexperts, and this, particularly for Darwin's *On the Origin of Species*, is an ingredient in their enduring power.

4. Anthony Gottlieb, "It Ain't Necessarily So," *New Yorker*, September 17, 2015, accessed June 29, 2015, http://www.newyorker.com/magazine/2012/09/17/it-aint-necessarily-so.

5. Darwin, *On the Origin of Species*.

CHAPTER 1. TWO MYSTERIES

1. Life's features are divided into two general categories, somatic and reproductive. Somatic features concern survival, while reproductive features of course concern the creation of new life. Since survival is not reproduction and reproduction is not survival, they exclude each other by designation. With the very significant exception of secondary sexual characteristics, such as those of sexual attraction (see section III), traits are either one or the other, not both. As I use the term *somatic*, it is not synonymous with *Soma*. *Soma* would probably be preferable, except that, regrettably, as the word is commonly used today, it includes all the features of the body including those of reproduction (except for germ cells).

2. Most high school biology or college Biology 101 textbooks explain the basic facts of reproduction. This includes the formation of germ cells and fertilization, and sometimes the major features of a selected species or two. There are many specialty textbooks that describe the

mechanisms of reproduction for various groups of organisms. Botany texts tell us about plant reproduction, for example: James D. Mauseth, *Botany: An Introduction to Plant Biology*, 5th ed. (Burlington, MA: Jones & Bartlett, 2012); Ray F. Evert and Susan E. Eichhorn, *Raven Biology of Plants*, 8th ed. (New York: W. H. Freeman, 2012); Spencer C. H. Barrett, ed., *Major Evolutionary Transitions in Flowering Plant Reproduction* (Chicago: University of Chicago Press, 2008); Sharman O'Neill and Jeremy A. Roberts, eds., *Plant Reproduction*, 1st ed. (DeKalb, IL: Blackwell, 2001); Hudson T. Hartmann, Dale Kester, Fred Davies, and Robert Geneve, *Plant Propagation: Principles and Practices*, 6th ed. (Prentice Hall, 1996). Invertebrate zoology texts outline the great diversity of reproductive mechanisms among invertebrate animals, for example, Edward E. Ruppert, Richard S. Fox, and Robert D. Barnes *Invertebrate Zoology: A Functional Evolutionary Approach*, 7th ed. (Cengage Learning, 2003); Ralph Buchsbaum, Mildred Buchsbaum, John Pearse, and Vicki Pearse, *Animals without Backbones: An Introduction to the Invertebrates* (Chicago: University of Chicago Press, 2013); Arthur C. Giese, ed., *Reproduction of Marine Invertebrates*, vols. 1–5 (New York: Academic, 1974–1980). Finally, comparative anatomy/physiology texts tell us about reproduction in vertebrates, for example: Kenneth Kardong, *Vertebrates: Comparative Anatomy, Function, Evolution*, 6th ed. (New York: McGraw-Hill, 2011); John E. Hall, *Guyton and Hall Textbook of Medical Physiology*, 12th ed. (Saunders, 2010); G. Prakash, *Reproductive Biology* (United Kingdom: Alpha Science, 2007); F. H. Bronson, *Mammalian Reproductive Biology* (Chicago: University of Chicago Press, 1991); Barrie, G. M. Jamieson, ed., *Reproductive Biology and Phylogeny of Birds, Part B: Sexual Selection, Behavior, Conservation, Embryology and Genetics* (Science Publishers, 2007); Barrie G. M. Jamieson, ed., *Reproductive Biology and Phylogeny of Fishes (Agnathans and Bony Fishes): Phylogeny, Reproductive System, Viviparity, Spermatozoa* (Science Publishers, 2009); Rámon Piñón, *Biology of Human Reproduction* (University Science Books, 2002); Richard E. Jones, *Human Reproductive Biology*, 3rd ed. (New York: Academic, 2006).

3. See G. C. Williams, *Sex and Evolution* (Princeton, NJ: Princeton University Press, 1975), John Maynard Smith, *The Evolution of Sex* (Cambridge: Cambridge University Press, 1978), and the recent Lucio Vinicius, *Modular Evolution* (Cambridge: Cambridge University Press, 2010).

CHAPTER 2. A MOST EXTRAORDINARY THEORY

1. Stephen Rothman, *Life beyond Molecules and Genes: How Our Adaptations Make Us Alive* (Philadelphia: Templeton, 2009).

2. Thomas S. Kuhn, *The Structure of Scientific Revolutions* (Chicago: University of Chicago Press, 1962).

3. Though Darwin's theory was initially controversial as a matter of science, the most passionate debate over the years has been religious and political. To get a sense of the political disputes at their fiery best, read the writings of Herbert Spencer in the nineteenth century and J. B. S. Haldane in the early- to mid-twentieth century, as well as books written about their ideas. To

find your way into their thoughts and the controversies, look at John Offer, ed., *Spencer: Political Writings* (Cambridge: Cambridge University Press, 1993); Richard Hofstadter, *Social Darwinism in American Thought* (University of Pennsylvania Press, 1945; repr., Beacon, 1992); Geoffrey M. Hodgson, "Social Darwinism in Anglophone Academic Journals: A Contribution to the History of the Term," *Journal of Historical Sociology* 17, no. 4 (2004): 428; Michael W. Taylor, *Men versus the State: Herbert Spencer and Late Victorian Individualism* (Oxford: Oxford University Press, 1992); J. B. S. Haldane, *Marxist Philosophy and the Sciences* (Random House,1939); Krishna R. Dronamraju, ed., *Haldane's Daedalus Revisited* (Oxford: Oxford University Press, 1995); J. B. S. Haldane, "The Last Judgment," *Possible Worlds and Other Essays* (London: Chatto & Windus, 1927); Krishna Dronamraju, ed., *What I Require from Life. Writings on Science and Life from J. B. S. Haldane* (Oxford: Oxford University Press, 2009); Mark B. Adams, "Last Judgment: The Visionary Biology of J. B. S. Haldane," *Journal of the History of Biology* 33 (2000): 457–91; Richard Jeffery, "C. S. Lewis and the Scientists," *The Chronicle of the Oxford University C. S. Lewis Society* 2, no. 2 (2005); J. B. S. Haldane, *Everything Has a History* (Allen & Unwin, 1951); J. B. S. Haldane, "A Dialectical Account of Evolution," *Science & Society* 1 (1937).

4. The word *physical* is meant to include all living things and all interactions among them. We are after all part of the "physical" world.

5. The term *survival of the fittest* was coined by the philosopher Herbert Spencer.

6. Though the political disputes about Darwinian theory have become muted in recent times, the religious arguments roil with seemingly unabated vehemence. Lately attention has centered on the concept of Intelligent Design. Opponents of Intelligent Design say that it is little more than warmed-over religious belief clothed in scientific garb, a modern reworking of nineteenth-century Natural Theology. For the ideas of Natural Theology, see William Paley's seminal *Natural Theology* (1802; repr., New York: Oxford University Press, 2006); Charles Bell, *The Hand. Its Mechanism and Vital Endowments as Evincing Design* (London: William Pickering, 1837; repr., Elibron Classics, 2004); Arthur O. Lovejoy, *The Great Chain of Being* (Cambridge, MA: Harvard University Press, 1936); E. S. Russell, *The Directiveness of Organic Activities* (Cambridge: Cambridge University Press, 1945); and Alan Olding, *Modern Biology and Natural Theology* (London: Routledge, 1991).

Adherents of Intelligent Design say that they are putting Darwinian theory to the fire of reason by pointing out its logical and existential weaknesses, inconsistencies, and fallacies. To get a sense of the debate, see Michael Denton, *Evolution: A Theory in Crisis*, 3rd ed. (Bethesda, MD: Adler and Adler, 1986); Phillip E. Johnson, *Darwin on Trial* (Downers Grove, IL: InterVarsity, 1993); Michael Denton, *Icons of Evolution: Science or Myth?* (Washington, DC: Regnery, 2002); John Wilson and William A. Dembski, *Uncommon Dissent: Intellectuals Who Find Darwinism Unconvincing* (Wilmington, DE: ISI Books, 2004); Michael Ruse, *Darwin and Design: Does Evolution Have a Purpose?* (Cambridge, MA: Harvard University Press, 2004); Jonathan Wells, *The Politically Incorrect Guide to Darwinism and Intelligent Design* (Washington, DC, Regnery, 2006); Michael Behe, *The Edge of Evolution* (New York: Free Press, 2007); William Dembski and Michael Ruse, eds., *Debating Darwin: From Darwin to DNA* (Cambridge: Cambridge University Press, 2007); U. Segerstrale, *Defenders of the Truth: The Battle for Science in the Sociobiology Debate and Beyond* (Oxford: Oxford University Press, 2000); and Stephen C. Meyer, *Darwin's*

Doubt: The Explosive Origin of Animal Life and the Case for Intelligent Design (HarperOne, 2013). Related volumes are David Stove, *Darwinian Fairytales* (New York: Encounter Books, 1995); Elliott Sober, *Evidence and Evolution* (Cambridge: Cambridge University Press, 2008); and David Berlinski, "The God of the Gaps," *Commentary* magazine, April 1, 2008.

7. Stephen Jay Gould, "Nonoverlapping Magisteria," *Natural History* 106 (March 1997): 16–22, accessed July 23, 2015, http://www.colorado.edu/physics/phys3000/phys3000_fa11/StevenJGoulldNOMA.pdf.

CHAPTER 3. A LOOK INSIDE DARWIN'S THEORY

1. The propositions that form the theory of evolution by means of natural selection were first presented in Charles Darwin, *On the Origin of Species by Means of Natural Selection* (London: John Murray, 1859) and in Charles Darwin and Alfred Wallace, "On the Tendency of Species to form Varieties and on the Perpetuation of Varieties and Species by Natural Means of Selection," *Proceedings of the Linnean Society* 3 (1858): 45–62. For a consideration of the theory almost a century later, see Ernst Mayr, *Systematics and the Origin of the Species* (New York: Columbia University Press, 1942).

2. A good place for a novice to start reading about evolution and Darwin's theory are the easy-to-read books of the late Stephen Jay Gould: *Ontogeny and Phylogeny* (Cambridge, MA: Harvard University Press, 1977); *Ever since Darwin* (New York: W. W. Norton, 1977); *Hen's Teeth and Horse's Toes* (New York: W. W. Norton, 1983); *An Urchin in the Storm* (New York: W. W. Norton, 1987); *Wonderful Life* (New York: W. W. Norton, 1989), and Richard Dawkins: *The Selfish Gene* (Oxford: Oxford University Press, 1976); *The Blind Watchmaker* (New York: W. W. Norton, 1986); *Climbing Mount Improbable* (New York: W. W. Norton, 1996); and *Unweaving the Rainbow* (Boston: Houghton Mifflin, 1998). Two recent books outlining the basis of evolution are Jerry A. Coyne, *Why Evolution Is True* (Penguin, 2009), and Donald R. Prothero, *Evolution—What the Fossils Say and Why It Matters* (New York: Columbia University Press, 2007).

3. See chap. 14 of this book and Stephen Rothman, *Life beyond Molecules and Genes: How Our Adaptations Make Us Alive* (Philadelphia: Templeton, 2009), chap. 7.

CHAPTER 4. THE MEANING OF NATURAL SELECTION

1. The full title of *On the Origin of Species* is *On the Origin of Species by Natural Selection or Preservation of Favoured Races in the Struggle for Life*. The word *races* is explosive today, but Darwin was referring to groups within species that display distinctive, favorable features (in relation to survival) compared to other groups of the same species. According to his theory, these differences were important factors in the separation of such groups from each other and led to new species.

2. Charles Darwin, *On the Origin of Species by Means of Natural Selection* (London: John Murray, 1859).

3. The concept of group selection has played an important if controversial role in the study of biological evolution. For an analysis, see Mark E. Borrello, *Evolutionary Restraints—The Contentious History of Group Selection* (Chicago: University of Chicago Press, 2010); Samir Okasha, *Evolution and the Levels of Selection* (Oxford: Oxford University Press, 2006); and Laurent Keller, ed., *Levels of Selection in Evolution* (Princeton, NJ: Princeton University Press, 1999).

4. Despite the theory's relatively rapid acceptance among scientists (it took about a decade after the publication of *On the Origin of Species* for most of Darwin's contemporaries to sign on), resistance to the theory continued for many years. For an account, see Peter J. Bowler, *The Eclipse of Darwinism* (Baltimore: Johns Hopkins University Press, 1983). The merging of evolutionary and genetic theories in the modern synthesis seemed to put an end to most scientific opposition.

5. For the initial exposition of the ideas of the modern synthesis and the related subject of population genetics, see the writings of its three cofounders: R. A. Fisher, *The Genetical Theory of Natural Selection* (Oxford: Oxford University Press, 1930); J. B. S. Haldane, *The Causes of Evolution* (London: Longmans, Green, 1932); Sewall Wright, "The Roles of Mutation, Inbreeding, Crossbreeding and Selection in Evolution," *Proceedings of the Sixth International Congress on Genetics* 1 (1932): 356–66; and Sewall Wright, *Evolution: Selected Papers* (Chicago: University of Chicago Press, 1986). Also see Theodosius Dobzhansky, *Genetics and the Origin of the Species* (New York: Columbia University Press, 1937).

6. Richard Dawkins, *The Ultraviolet Garden*, Royal Institute Christmas Lecture, no. 4, 1991.

7. Subsequent to the development of the modern synthesis, the criticism of natural selection that remained focused in the main on its consequences, rather than on natural selection itself. Either one believed, like Wallace, that all of life's evolved features are the products of natural selection and consequently are adaptations (this became known as Adaptationism) or questioned the ubiquity of adaptations.

For the Adaptationist perspective, see George C. Williams, *Adaptation and Natural Selection* (Princeton, NJ: Princeton University Press, 1966); Verne Grant, *The Origin of Adaptations* (New York: Columbia University Press, 1963); Marjorie Grene, ed., *Dimensions of Darwinism: Themes and Counter-themes in Twentieth-Century Evolutionary Theory* (Cambridge: Cambridge University Press, 1983); Niles Eldredge, *Unfinished Synthesis: Biological Hierarchies and Modern Evolutionary Thought* (Oxford: Oxford University Press, 1985) and *Fossils: The Evolution and Extinction of Species* (Princeton, NJ: Princeton University Press, 1991); M. R. Rose and G. V. Lauder, eds., *Adaptation* (New York: Academic, 1996); and Steven Hecht Orzack and Elliot Sober, eds., *Adaptationism and Optimality* (Cambridge: Cambridge University Press, 2001). For Wallace's perspective, see Andrew Berry, *Infinite Tropics* (London: Verso, 2002).

For questioning the Adaptationist perspective, see S. J. Gould and R. C. Lewontin, "The Spandrels of San Marco and the Panglossian Paradigm: A Critique of the Adaptationist Programme," *Proceedings of the Royal Society of London, B* 205 (1971): 581–98; Stephen Jay Gould, "The Exaptive Excellence of Spandrels as a Term and Prototype," *Proceedings of the National Academy of Sciences (USA)* 94 (1997): 10750–55; R. C. Lewontin, *Biology as Ideology: The Doc-*

trine of DNA (New York: HarperCollins, 1993); Elliot Sober, "Six Sayings about Adaptationism," in *The Philosophy of Biology*, ed. David Hull and Michael Ruse (Oxford: Oxford University Press, 1998); and Stephen Jay Gould, *The Structure of Evolutionary Theory* (Cambridge, MA: Harvard University Press, 2002).

8. Over the past half century, the challenge to natural selection has become frontal. In the vanguard has been evolutionary genetics, the modern incarnation of population genetics. As it developed, evolutionary genetics came to relegate natural selection to a subordinate role in evolution, placing a greater emphasis on mutations, variety, and, most importantly, reproduction. The genetic study of populations of organisms, begun by Wright and Fisher in the '30s, was based in great part on Wright's concept of genetic drift. Genetic drift is random changes in the genetic composition of groups that result from reproduction. It has much in common with the ideas about reproductive evolution presented in section V of this book. Though neither Wright nor Fisher saw genetic drift as a major element in evolution, in 1968 Motoo Kimura proposed in "Evolutionary Rate at the Molecular Level" (*Nature* 217 [1968]: 624–26), that neutral genetic shifts similar to those of genetic drift are central.

9. Richard Lewontin, "Not So Natural Selection," *New York Review of Books*, May 27, 2010, pp. 34–36.

10. To explore the ideas of evolutionary genetics, see John Maynard Smith, *Evolutionary Genetics* (Oxford: Oxford University Press, 1998) and Richard C. Lewontin, *The Genetic Basis of Evolutionary Change* (Columbia, NJ: Columbia University Press, 1974), as well as a recent analysis of natural selection and genetic drift by Elliott Sober, *Evidence and Evolution: The Logic behind the Science* (Cambridge: Cambridge University Press, 2008).

CHAPTER 5. ONE FOR ALL

1. From the time that Darwin introduced natural selection as the agency of biological evolution, it has been understood to apply to the evolution of the reproductive features of life as well as its somatic features. Attempts to account for reproductive evolution in terms of natural selection have followed one or another of the three paths discussed here and in the following chapters. Allusions to them can be found scattered throughout the literature on evolution (for example, see the references listed for the prologue as well as chapters 3 and 4).

2. The indispensable event of reproduction is molecular duplication, and its origin is shrouded in the mystery of an unknown, and probably unknowable, occurrence in the long-distant past. The most popular view today is that reproduction began as straightforward chemistry prior to the emergence of the cell, indeed prior to the emergence of life. Simple chemical precursors to DNA, perhaps including structurally simple RNA molecules (see Leslie Orgel, "The Origin of Life on Earth," *Scientific American* 271, no. 81 [1994]) duplicated themselves. But whenever and however it happened, the origin of molecular duplication and consequently the origin of reproduction is, as said, by all accounts a great oddity, the result of a singularity, something that happened just once some four billion years ago. For the first exposition of modern ideas

about the beginnings of reproduction and of life, see A. I. Oparin's classic *Origin of Life* (1938; Dover, 2003). Also, John Maynard Smith and Eörs Szathmáry, *The Origins of Life: From the Birth of Life to the Origin of Language* (Oxford: Oxford University Press, 1999).

CHAPTER 6. THE OTHER BEGINNING

1. Some believe that molecular duplication arose subsequent to the appearance of the cell. See Harold J. Morowitz, *Beginnings of Cellular Life; Metabolism Recapitulates Biogenesis* (New Haven, CT: Yale University Press, 1992)].

CHAPTER 7. BEYOND REPLICATING MOLECULES AND REPRODUCING CELLS

1. To the best of my knowledge there is not a single modern book, book chapter, or even an article that both accepts natural selection *and* questions its role in the evolution of the reproductive features of life.

2. Karl Popper, *The Logic of Scientific Discovery* (Abingdon, UK, and New York: Routledge, 2002).

3. Jerry A. Coyne in *Why Evolution Is True* (London: Penguin, 2009), reflecting on what appears to be the common view, tells us that sexual selection "is just a subset of natural selection."

CHAPTER 8. IRIDESCENT BRILLIANCE

1. His reasons can be found in his *Contributions to the Theory of Natural Selection* (New York: Macmillan, 1870) and recently in Andrew Berry, *Infinite Tropics: An Alfred Russel Wallace Anthology* (London: Verso, 2002).

2. Helena Cronin's *The Ant and the Peacock: Altruism and Sexual Selection from Darwin to Today* (Cambridge: Cambridge University Press, 1991) provides the first modern accounting of Darwin's theory of sexual selection. Her discussion does not question the idea that the sexual features of life evolved as the consequence of natural selection.

3. One problem that the evolution of features of sexual attraction and the process of sexual selection raises for the theory of natural selection is the need for synchronicity. That is, as the attractant evolves, so must the response to it, synchronously. This is because sexual selection can only occur when the attractant is displayed and attracted to simultaneously. Otherwise, what would cause an attractant to evolve in one sex before the means of it being seen as such by the other existed? Or the other way around, what sense does it make to be attracted to something that does not yet exist?

How does Darwinian theory explain this requirement, given that the attractant and the response are expressed in different organisms, male and female, each with a unique evolutionary history in regard to reproduction? There have been two approaches. Wallace offered one and Darwin the other. Wallace proposed that the attractant and being attracted to it evolved separately for purposes of *survival*, not reproduction, and only later became linked to sex. Consider legs in humans. Bone length and placement, muscle form, mass, and the movement of it all, serve the somatic purposes of locomotion, the purposes of survival, but at some point in evolutionary history, certain aspects or embodiments of the human leg came to be seen as sexy. The same can be said of all sorts of anatomical features, everything from facial expression (for instance, it may indicate fearsomeness or obsequiousness as well as beauty) to hair standing on end (to reduce heat loss or display fright or impending aggression, as well as sexual interest). What serves somatic purposes in one situation may serve sexual ones in another.

We can even imagine that the plumage of the peacock evolved for some somatic purpose, say, its shimmery quality frightened off predators, or in a faint light the eyes of its feathers looked like so many animals arrayed against the attacker. Still, however limited or fulsome our imagination, according to this view there is some somatic purpose, a purpose related to survival, for every sexually imbued feature. This of course leaves the extent of our understanding to our imagination, not science. Even so, it does not tell us how particular somatic features became objects of sexual desire, when other incarnations of the same feature, often not significantly different materially or functionally, did not.

Darwin's explanation was simpler in that it did not require some unknown means to suffuse a feature with sexual meaning. In his view, the choice of a partner memorialized that meaning. Reproduction takes place only if the feature is attractive (not a synonym of *beautiful* but an adjectival form of the present participle, *attracting*) relative to another, and of course only with reproduction do things evolve. Through enhanced attractiveness and deepening desire, reproduction became increasingly likely, increasingly desirable. Though this still does not explain why particular features became sexually attractive, nonetheless both attractant and response were optimized as they evolved in tandem across generations.

CHAPTER 10. IN THE EYES OF THE BEHOLDER

1. To explore concepts of beauty, see Roger Scruton, *Beauty: A Very Short Introduction* (Oxford: Oxford University Press, 2009); Umberto Eco, ed., *History of Beauty* (New York: Rizzoli, 2010); and Umberto Eco, ed., *On Ugliness* (New York: Rizzoli, 2007).

CHAPTER 11. A GREAT CONTRADICTION

1. To learn about Aristotle's four causes, see Jonathan Barnes, ed., *The Complete Works of Aristotle, The Revised Oxford Translation* (Princeton, NJ: Princeton University Press, 1984).

2. Francis Bacon, *Novum Organum Scientiarum*, 1620; *The New Organon*, ed. Lisa Jardine and Michael Silverthorne (Cambridge: Cambridge University Press, 2000).

3. For Aristotle's ultimate or final cause, see *Parts of Animals* I.1 or *Physics* II.5-9. For a discussion of its controversial history, see Monte Ransome Johnson, *Aristotle on Teleology* (Oxford: Oxford University Press, 2005); Mariska Leunissen, *Explanation and Teleology in Aristotle's Science of Nature* (Cambridge: Cambridge University Press, 2010); James G. Lennox and Robert Bolton, *Being, Nature, and Life in Aristotle* (Cambridge: Cambridge University Press, 2010).

4. The idea of purpose and final causes served as the basis for teleology. See Francisco Ayala, *Nature's Purposes: Analyses of Function and Design in Biology* (Cambridge, MA: MIT Press, 1998); E. S. Russell, *The Directiveness of Organic Activities* (Cambridge: Cambridge University Press, 1945); and Andrew Woodfield, *Teleology* (Cambridge: Cambridge University Press, 1976). For the disputatious history of teleological thinking in the biological sciences, see Ernst Mayr, "The Idea of Teleology," *Journal of the History of Ideas* 53 (1992): 117–35.

5. As we consider the idea of purpose, we have to be careful about the meaning of words. I used the word *choice* in reference to nature, "nature's choice." Likewise, in the following sentence, I referred to "natural selection guiding events." To this point you may not have noticed their use, but now they may give you the impression that I am saying that nature has a purpose, that it chooses and guides with some sort of intention.

But *choice* is merely a synonym for *selection*, as in *natural selection*. Similarly, to say that nature is guiding events is to say no more than that it influences a course of action. These words are not meant to imply intentionality, any more than Darwin's use of the term *natural selection* did.

The concept of natural selection is that when confronted with environmental challenge some organisms, some mutations do better than others. The comparative element is critical. Without choice, without selection between organisms, without nature's unintentional guidance, there would be no evolution. Natural selection is an agency of choosing, of guidance in the absence of intention.

CHAPTER 12. PURPOSE

1. That is, unless you imagine, for example, that cars just happen to be means of transportation, a kind of mutation, chance occurrences. And even then, what purpose would it serve for us to imagine that they serve no purpose?

CHAPTER 13. DARWIN'S TELEOLOGY OF SURVIVAL

1. See James Lennox, "Darwin Was a Teleologist," *Biology and Philosophy* 8, no. 4 (1993): 409–21; Stephen C. Meyer, *Darwin's Doubt: The Explosive Origin of Animal Life and the Case for Intelligent Design* (HarperOne, 2013).

2. Charles Darwin, *The Life and Letters of Charles Darwin*, ed. Francis Darwin (New York: Basic Books, 1959), p. 367.

3. You might say, "But survival is an instinct, not a purpose." The organism has no choice but to try to survive, and how can there be purpose in that? But then why does it have an instinct to survive, to what end? Objects in the inanimate world don't have it, nor does the genetic code, code for it. Yet life seeks to persevere, to preserve itself. It is not indifferent to its fate. The impulse to survive is both a sign of life and an indicator of its purpose. It is the purpose of each and every adaptive feature of living things and of them all combined. Calling it an instinct may make it seem more mechanical, but as we shall see purposes can be mechanical. Also, there are many fascinating exceptions to this "instinct." For instance, there is the lemmings' plunge into the sea and the male honeybees' drive to copulate and die as a consequence.

4. Stephen Rothman, *Lessons from the Living Cell* (New York: McGraw-Hill, 2002).

5. Ibid.

6. Stephen Rothman, *Life beyond Molecules and Genes: How Our Adaptations Make Us Alive* (Philadelphia: Templeton, 2009).

7. The circumstances of biological chemicals and structures before life itself emerged were probably relatively pacific, and the danger minimal. Biological evolution began in earnest when scarcity became the rule and with it predation and the need to seize the resources of life in order to survive. See chapters 5 and 6, and n. 4 above.

CHAPTER 14. THE LAWS OF NATURE

1. How, you might ask, can compliance with physical law serve a purpose? But of course we comply with all sorts of laws with express purposes. For instance, we comply with traffic laws with the purpose of avoiding fines. Still, aren't physical laws different? Aren't they are obligatory? We have no choice but to comply. I can avoid the need to comply with traffic laws simply by deciding not to drive, but such an option is not available in regard to physical law. It is not in my hands to avoid the need to comply.

However, as it turns out, compliance *is* a matter of choice when we consider it for biological adaptations. Fur, for example, protects mammals that live in cold climates, such as polar bears. Of course most living things lack fur and hence need not comply with the laws of heat transfer as they are related to the *purpose of fur*. Plants do not have fur, nor do they have feathers for flying, gills for swimming, or legs for walking. Absent these features and their ends, they do not have to comply with physical laws as they relate to them. They are not applicable.

But isn't the purpose of compliance the same as that of the adaptation, survival? Isn't there just this one purpose, and it is evinced by the adaptation's fealty to physical law? But survival and compliance are not the same thing. To comply is not to survive, and vice versa. They are different purposes, both of which are required for the adaptive function to exist.

2. For the application of the laws of heat transfer to living things, see Walter B. Cannon, *Wisdom of the Body* (W. W. Norton, 1932, rev. ed., 1963) and Stephen Rothman, *Life beyond Molecules and Genes: How Our Adaptations Make Us Alive* (Philadelphia: Templeton, 2009). For a general physical discussion of heat transfer, see Richard P. Feynman, Robert B. Leighton, and Matthew Sands, *The Feynman Lectures on Physics: Mainly Mechanics, Radiation, and Heat* (New York: Basic Books, 2011), vol. 1. Heat is a measure of atomic and molecular movement and as such is a measure of density. All of life's adaptations, even those that have nothing ostensibly to do with temperature, evolved with absolute fidelity to its demands. It has influenced the evolution of everything about life from molecules to organisms, from the rate and character of chemical reactions, to our anatomy, to our shape and structure, to basic physical processes such as movement, encompassing each and every physiological and chemical aspect of life without exception.

3. Needless to say, temperature is not alone in this. Every characteristic of living things displays absolute fidelity to each and every physical, chemical, and mathematical law. The laws of the physical world played a central role in the evolution of our adaptive character. Take the falling rock. Certainly the rock has no influence on life simply because it exists, and yet as explained when it falls it can be dangerous to any living thing in its path. Depending upon your size and that of the rock, it may injure or even crush you. This danger, a function not only of the rock's size but also of its mass and density, is the final cause of its action. It inheres to it as the result of gravity (the reason why it is falling). When the inert rock takes on a kinetic dimension it can become dangerous, and its physical properties present a danger to living things.

Opposing this danger are adaptations in the living thing designed to notice and dodge the rock. Just as the rock poses a danger to us because of its physical characteristics, adaptations to that danger are also expressions of physical law. For example, I may perceive photons reflected off the rock (I see the rock), and from this I assess the position of its accelerating mass in space, estimate its size and density, and finally gauge the danger. If I determine that there is a real threat, then using the laws of mechanics, I engage my various muscles and joints to accelerate my mass to a velocity sufficient to dart out of the rock's path, letting it fall harmlessly to the ground.

4. Rothman, *Life beyond Molecules and Genes.*

5. Ibid.

6. For a contemporary discussion of the heart's raison d'être and its physical basis, see ibid. For a general discussion of the heart and cardiovascular system, see any physiology text, such as John E. Hall, *Guyton and Hall Textbook of Medical Physiology*, 12th ed. (Saunders, 2010) or specialty texts devoted to the subject, for example, David E. Mohrman and Lois Jane Heller *Cardiovascular Physiology*, 5th ed. (New York: McGraw-Hill/Appleton & Lange, 2002).

7. Gravity's influence is not merely a matter of falling bodies. It is all-embracing. For instance, vertebrates are not vertebrates by accident. Animals of their size could not exist without a bony vertebral column and calcified long bones or some equivalent structures to counter gravita-

tional forces. Internally, there is a web of connective tissues, masses of crisscrossing collagen fibers, ligaments and tendons that suspend and cushion our internal organs and prevent their gravitational collapse. Feathers, wings, light torsos, and body shape evolved in birds and insects to help them overcome gravity's downward force and become airborne.

Even the heart is a slave to gravity. The heart of each and every species with a circulatory system evolved to pump blood at just the right pressure to ensure perfusion of organs and tissues above the heart (the head in humans), and the return of blood to it from organs and tissues below the heart, closer to the ground, like our lower limbs, all of it to overcome gravity's incessant downward pull.

We also see the demands of gravity in aging. The reduced length of telomeres and number of mitochondria aside, the decrepitude of old age has much to do with the inexorable ravages of gravity. Despite the heroic resistance of our bodies, gravity works without pause to pull us to the ground, to make us one with the planet. We shrink as we grow old; our vertebrae become compressed, often producing limb and back pain; and our knees no longer provide the cushion they once did separating upper and lower leg bones. These effects cascade. Problems with ambulation result in decreased conditioning, and that increases our susceptibility to cardiovascular disease. Though the effects are gradual and in their early stages barely noticeable, gravity is a harsh and relentless master. In any event, whatever the feature and the physical constraint, gravity included, the laws of physics and the rules of mathematics played a fundamental, controlling and indeed an all-encompassing role in life's evolution.

8. Beginning about a half-century ago, some physical scientists claimed that life's final cause could be found not in the mundane events of life on earth but in nothing less than the evolution of the cosmos itself. According to this point of view, after the big bang cosmic evolution tended inexorably toward life's emergence. Known as the anthropic cosmological principle or sometimes the Goldilocks enigma, it argues that life was fated by the physical circumstances of the universe (at least our universe) as it expanded and evolved.

Preordained by the circumstances of cosmic evolution, by its rates, constants, and mass, life emerged. The thinking goes that if things had been slightly different, if the cosmos had evolved a little more slowly or a little more quickly, if the mass of the universe had been a little larger or a little smaller, the physical constants just a little different, a little higher or lower, life would not have been possible. Whatever the reasons, the universe evolved to provide exactly the conditions needed for life's emergence, and in so doing served as its final cause. In much the same way that gravity affects the heart's properties, life was a matter of physical circumstance.

There have been three reactions to this idea. Probably the most common among biologists has simply been to dismiss it. Evolutionary biologist Stephen Jay Gould wondered pointedly whether the anthropic cosmological principle doesn't get it completely backward. Isn't it far more likely, as presented here, that life arose to comply with the particular properties of the universe, than the other way around? Those who found the idea congenial expressed support for it in two diametrically opposite ways. For some the concept provided evidence for a purely physical basis for life's origin and evolution, while others saw it as proof of God's directing hand.

But either way, even if the physical world provided the necessary preconditions for life's emergence, however salubrious, they were not enough for life to simply arise. Prerequisites, enabling conditions, however beneficial, yield nothing in the absence of accompanying sufficient conditions. As discussed, a satisfactory account of life's emergence must explain its sufficient as well as its necessary conditions; that is, it must explain the basis for its materialization from them. To say that a carbon-based *system* of some sort was predestined by the events that followed the big bang is not the same as saying that carbon-based *life* was similarly fated.

It is sometimes said, rather glibly, I believe, that this is not a problem because life is a self-assembling physicochemical system, and that given the necessary preconditions it will emerge spontaneously. But life did not simply gather itself up because the potential to do so was there. Anthropic principle or not, without natural selection and its adaptations and critically without reproduction to give natural selection evolutionary meaning, life as we know and understand it could not have come into being . However promising the circumstances, life could not have materialized in the absence of its sufficient causes.

As for selection in regard to reproduction, it was not merely critical for the evolution of the reproductive features of life; it was indispensable for the evolution of life in toto. Though natural selection fits comfortably into the cosmological setting just described, reproductive selection does not. In the presence of life's various antecedents and given propitious physical conditions, natural selection would occur in and by itself. Life would have come about, well, naturally; that is, automatically, if not exactly spontaneously. In the face of danger, adaptive advantage would arise (see Rothman, *Life beyond Molecules and Genes*), and natural selection would impose its choices. The problem is that however automatic, natural selection could only have had evolutionary meaning in the presence of reproduction, and to the best of our knowledge there is no feature of the natural world outside of life itself (actually death) that predisposes to its occurrence. As said, molecular duplication, its basis and origin, is thought to have been not only an uncommon occurrence but also singular and fortuitous, rare and accidental.

To learn about the anthropic cosmological principle, see R. H. Dicke, "Dirac's Cosmology and Mach's Principle," *Nature* 192 (1961): 440–41; B. Carter, "Large Number Coincidences and the Anthropic Principle in Cosmology," *IAU Symposium 63: Confrontation of Cosmological Theories with Observational Data* (Dordrecht, Holland: Reidel, 1974); John D. Barrow and Frank J. Tipler, *The Anthropic Cosmological Principle* (Oxford: Oxford University Press, 1988); Paul Davies, *The Goldilocks Enigma: Why Is the Universe Just Right for Life?* (Houghton Mifflin Harcourt, 2008); William Lane Craig, "The Anthropic Cosmological Principle," *International Philosophical Quarterly* 27 (1987): 437–47; M. A. Walker and M. M. Cirkovic, "Anthropic Reasoning, Naturalism and the Contemporary Design Argument," *International Studies in the Philosophy of Science* 20 (2006): 285–307; Victor J. Stenger, "Anthropic Design," *Skeptical Inquirer* 23 (1999): 40–43; Jesús Mosterín, "Anthropic Explanations in Cosmology," in ed. P. Háyek, L. Valdés, and D. Westerstahl, *Proceedings of the 12th International Congress of Logic, Methodology and Philosophy of Science* (London: King's College, 2005), pp. 441–73; Stuart Ross Taylor, *Destiny or Chance: Our Solar System and Its Place in the Cosmos* (Cambridge: Cambridge University Press, 1998); Max Tegmark, "On the Dimensionality of Spacetime," *Classical and Quantum Gravity* 14 (1997): L69–L75; Marcus Chown, "Anything Goes," *New Scientist*, June 6, 1998.

CHAPTER 16. THE MEANS

1. In a letter to the editor of the *Wall Street Journal* (Dick Gruber, May 12, 2013) it was pointed out that sooner or later the monkey would press Ctrl-A or Select All followed by Delete and would have to start all over again.

2. See chap. 14, n. 1.

3. Even on its own terms, the division of life into somatic and reproductive features suggests the need for a separate reproductive evolution. After all, the concerns of the two—survival of the existing and creation of the new—are entirely different.

4. The idea of a reproductive selection has its roots in evolutionary or population genetics (chapter 4) as well as in Darwin's theory of sexual selection (chapters 8–10). In regard to the first, reproduction, rather than natural selection is seen as the central driving force of biological evolution. In the second, Darwin introduces a means of selection in addition to natural selection based on reproductive traits.

5. The book was originally published in 1862 under the explanatory title *On the Various Contrivances by Which British and Foreign Orchids Are Fertilised by Insects, and on the Good Effects of Intercrossing* (London: John Murray, 1877).

CHAPTER 17. THE FINAL CAUSE OF REPRODUCTION

1. Darwin's theory did not jettison final causes, any more than it relied on a transcendent God. What it did was bring them back to earth, to a simple quotidian engagement with the environment.

2. See chapter 14, particularly note 8.

CHAPTER 19. ON CHAIRS AND CAROB TREES

1. For a consideration of the relation between science and religion, see Stephen Jay Gould's *Rocks of Ages: Science and Religion in the Fullness of Life* (New York: Ballantine, 1999). Three relatively recent books by atheists are Daniel Dennett's *Breaking the Spell: Religion as a Natural Phenomenon* (New York: Penguin, 2007); Christopher Hitchens's *The Portable Atheist: Essential Readings for the Nonbeliever* (Cambridge, MA: Da Capo, 2007); and Richard Dawkins's *The God Delusion* (London: Bantam, 2006).

For the agnostic perspective, see Bertrand Russell *Am I an Atheist or an Agnostic?* (Girard, KS: Haldeman-Julius, 1949); Bertrand Russell, *Why I Am Not a Christian and Other Essays on Religion and Related Subjects*, ed. Paul Edwards (New York: Touchstone, 1967); Thomas Henry Huxley *Man's Place in Nature: Man's Place in Nature and Other Anthropological Essays* (Whitefish, MT: Kessinger, 2005), as well as David Hume's *Concerning Natural Religion*, ed. Richard Popkin (Indianapolis: Hackett, 1998).

For the philosophical and religious skeptic, there is Panayot Butchvarov, *Skepticism about the External World* (New York: Oxford University Press, 1998). Also, look at Karl Giberson, *Worlds Apart: The Unholy War between Religion and Science* (Kansas City, KA: Beacon Hill, 1993); Gary Ferngren, ed. *Science and Religion: A Historical Introduction* (Baltimore, MD: Johns Hopkins Press, 2002); Paul Kurtz, Barry Karr, and Ranit Sandhu, *Science and Religion: Are They Compatible?* (Amherst, NY: Prometheus Books, 2003); Ken Wilber, *The Marriage of Sense and Soul: Integrating Science and Religion* (New York: Broadway, 1999); and Kenneth R. Miller, *Finding Darwin's God: A Scientist's Search for Common Ground between God and Evolution* (New York: Harper Perennial, 2007).

2. This story may be apocryphal, but whatever Twain's motivation, he did throw some of his characters down a well to get rid of them. See Author's Note to "Those Extraordinary Twins."

CODA

1. Stephen Rothman, *Lessons from the Living Cell: The Limits of Reductionism* (New York: McGraw-Hill, 2002).
2. Karl Popper, *The Logic of Scientific Discovery*, 1934; English transl. by author, 2d ed. (Abingdon, UK: Routledge, 2002).
3. Rothman, *Lessons from the Living Cell.*
4. Thomas S. Kuhn, *The Structure of Scientific Revolutions* (Chicago: University of Chicago Press, 1962).
5. Ibid.
6. Ibid.
7. Rothman, *Lessons from the Living Cell.*
8. Stephen Rothman, *Life beyond Molecules and Genes: How Our Adaptations Make Us Alive* (Philadelphia: Templeton, 2009).
9. Ibid.
10. Robert Rosen, "What Is Biology?," *Computers & Chemistry* 18, no. 3 (1994): 347–52.
11. Rothman, *Life beyond Molecules and Genes.*
12. Elizabeth Pennisi, "Finally the Book of Life and Instructions for Navigating It," *Science* 288 (2000): 2304–2307.
13. Rothman, *Lessons from the Living Cell.*
14. Rothman, *Life beyond Molecules and Genes.*

INDEX